PEER-LED TEAM LEARNING

GENERAL CHEMISTRY

SECOND EDITION

David K. Gosser
City College of New York

Victor S. Strozak
City University of New York

Mark S. Cracolice
University of Montana, Missoula

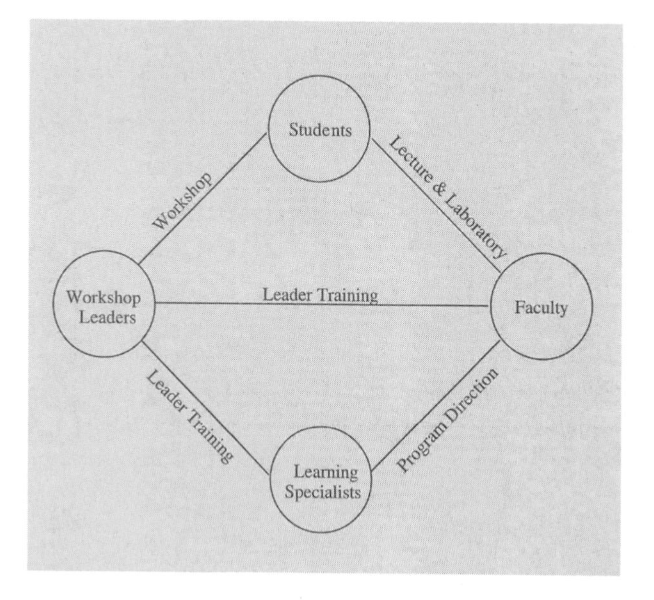

ei PRENTICE HALL SERIES IN EDUCATIONAL INNOVATION

PEARSON
Prentice Hall
Upper Saddle River, New Jersey 07458

Project Manager: *Kristen Kaiser*
Production Editor: *Donna Young*
Assistant Managing Editor: *Beth Sweeten*
Executive Managing Editor: *Kathleen Schiaparelli*
Manufacturing Buyer: *Alan Fischer*
Copy Editor: *Brian I. Baker*
Art Director: *Jayne Conte*
Manager, Cover Visual Research & Permissions: *Karen Sanatar*
Cover Design: *Bruce Kenselaar*

Cover Images
top: Getty Images, Inc.; bottom right: Gary Buss / Taxi / Getty Images, Inc.;
bottom left: Nick Rowe / Photodisc Green / Getty Images, Inc.;
bottom center: Getty Images, Inc.; bottom right: Gary Buss / Taxi / Getty Images, Inc.

Pearson Prentice Hall
Pearson Education, Inc.
Upper Saddle River, NJ 07458

Printed in the United States of America
10 9 8 7 6 5 4 3

ISBN 0-13-146444-2

Pearson Education LTD., *London*
Pearson Education Australia PTY. Limited, *Sydney*
Pearson Education Singapore, Pte. Ltd.
Pearson Education North Asia Ltd, *Hong Kong*
Pearson Education Canada, Ltd., *Toronto*
Pearson Educacion de Mexico, S.A. de C.V.
Pearson Education—Japan, *Tokyo*
Pearson Education Malaysia, Pte. Ltd.

Preface

The second edition of *Peer-Led Team Learning: General Chemistry* maintains the underlying philosophy and approach of the first edition, that is, active learning in peer-led groups engages students in the process of learning chemistry. This engagement results in improved understanding of chemistry concepts and of the process of science. The Peer-Led Team Learning model also helps you develop the communication and teamwork skills that are critical in the 21st-century workplace.

The second edition is also designed to be used as a workbook. It is purposely the size of a notebook in the hope that you, the chemistry student, will carry it around and use it frequently. It has ample space for you to record your notes and questions about the material. And, most importantly, it includes space to record the results of group work—the solutions to the workshop problems and the thinking behind those solutions. The workshop problems reflect the breadth of scientific problem solving—namely, the use of flowcharts, model building, simulations, and estimations—to help you understand and apply chemistry concepts. Week by week, you will be creating your own solutions guide that will be full of your thoughts and reflections on the problems you've solved. It will have a record of the logic that you and your classmates used to figure out the answers and the process you used to search out the solutions. You can refer to your solutions guide throughout the semester just as you refer to your lecture notes. You can use it as a study guide for examinations and a reference for subsequent chemistry courses. This is a resource that you will keep long after you've traded in your used textbook.

The changes in the second edition were inspired by feedback from reviewers and the students and faculty who have used the first edition over the past five years. We have adopted their suggestions to make the book more user friendly and to simplify the language and writing style. We have carefully examined the entire text, rewriting when necessary, to ensure that all content is presented in a theoretically accurate and pedagogically sound manner. By way of example, we have provided an improved and conceptually more correct treatment of enthalpy of formation in Unit 7 and have introduced the correct treatment for the kinetics of the steady state in Unit 17. Additionally, the Unit 17 workshop now includes new simulation parameters that demonstrate that a steady state requires neither a slow first step nor an intermediate of particularly low concentration. We have divided the text into smaller, bite-sized pieces that relate directly to the problems in the self-test and the workshop. We have changed the self-test format from a single test at the end of the unit to several smaller tests that follow each bite-sized section of text. We hope that this new configuration provides opportunities to think about the concept and to self-assess understanding before proceeding further.

The book's 24 chapters (workbook units) parallel the sequence of a typical general chemistry textbook. Each unit ends with a workshop that has more problems than can be done in a single group meeting. This was deliberately done to give instructors the ability to choose the problems that best meet their local curriculum needs. Each workshop includes numerous suggestions to facilitate collaborative work on the problems. While the workbook is tailored for a Peer-Led Team Learning environment, it can also be used as a supplement in those traditional lecture courses in which instructors encourage collaborative learning.

We encourage anyone who has comments or feedback about the text or interest in Peer-Led Team Learning to contact the authors.

We want to thank all the individuals who have made contributions to the workbook over the past few years. We are grateful to all of the students and group leaders who have made contributions throughout the life of the Workshop Project. We are particularly grateful to the National Science Foundation, Division of Undergraduate Education, for its support for the Workshop Project through the Chemistry Systemic Change Initiative and National Dissemination Program. For more information about Peer-Led Team Learning and the Workshop Project, visit the project's official website at www.pltl.org.

Finally, we would like to express our appreciation to Pearson Prentice Hall for their generous support of the dissemination of this text as part of their educational innovation program.

D. K. Gosser (gosser@sci.ccny.cuny.edu)

V. S. Strozak (vstrozak@gc.cuny.edu)

M. S. Cracolice (mark.cracolice@umontana.edu)

New York, NY and Missoula, MT

Period	Main groups 1A[a] 1	 2A 2	Transition metals 3B 3	 4B 4	 5B 5	 6B 6	 7B 7	 8B 8	 8B 9	 8B 10	 1B 11	 2B 12	Main groups 3A 13	 4A 14	 5A 15	 6A 16	 7A 17	 8A 18
1	1 **H** 1.00794																	2 **He** 4.002602
2	3 **Li** 6.941	4 **Be** 9.012182											5 **B** 10.811	6 **C** 12.0107	7 **N** 14.0067	8 **O** 15.9994	9 **F** 18.998403	10 **Ne** 20.1797
3	11 **Na** 22.989770	12 **Mg** 24.3050											13 **Al** 26.981538	14 **Si** 28.0855	15 **P** 30.973761	16 **S** 32.065	17 **Cl** 35.453	18 **Ar** 39.948
4	19 **K** 39.0983	20 **Ca** 40.078	21 **Sc** 44.955910	22 **Ti** 47.867	23 **V** 50.9415	24 **Cr** 51.9961	25 **Mn** 54.938049	26 **Fe** 55.845	27 **Co** 58.933200	28 **Ni** 58.6934	29 **Cu** 63.546	30 **Zn** 65.39	31 **Ga** 69.723	32 **Ge** 72.64	33 **As** 74.92160	34 **Se** 78.96	35 **Br** 79.904	36 **Kr** 83.80
5	37 **Rb** 85.4678	38 **Sr** 87.62	39 **Y** 88.90585	40 **Zr** 91.224	41 **Nb** 92.90638	42 **Mo** 95.94	43 **Tc** [98]	44 **Ru** 101.07	45 **Rh** 102.90550	46 **Pd** 106.42	47 **Ag** 107.8682	48 **Cd** 112.411	49 **In** 114.818	50 **Sn** 118.710	51 **Sb** 121.760	52 **Te** 127.60	53 **I** 126.90447	54 **Xe** 131.293
6	55 **Cs** 132.90545	56 **Ba** 137.327	71 **Lu** 174.967	72 **Hf** 178.49	73 **Ta** 180.9479	74 **W** 183.84	75 **Re** 186.207	76 **Os** 190.23	77 **Ir** 192.217	78 **Pt** 195.078	79 **Au** 196.96655	80 **Hg** 200.59	81 **Tl** 204.3833	82 **Pb** 207.2	83 **Bi** 208.98038	84 **Po** [208.98]	85 **At** [209.99]	86 **Rn** [222.02]
7	87 **Fr** [223.02]	88 **Ra** [226.03]	103 **Lr** [262.11]	104 **Rf** [261.11]	105 **Db** [262.11]	106 **Sg** [266.12]	107 **Bh** [264.12]	108 **Hs** [269.13]	109 **Mt** [268.14]	110 [271.15]	111 [272.15]	112 [277]	113 [284]	114 [289]	115 [288]	116 [292]		

Series														
*Lanthanide series	57 ***La** 138.9055	58 **Ce** 140.116	59 **Pr** 140.90765	60 **Nd** 144.24	61 **Pm** [145]	62 **Sm** 150.36	63 **Eu** 151.964	64 **Gd** 157.25	65 **Tb** 158.92534	66 **Dy** 162.50	67 **Ho** 164.93032	68 **Er** 167.259	69 **Tm** 168.93421	70 **Yb** 173.04
†Actinide series	89 **†Ac** [227.03]	90 **Th** 232.0381	91 **Pa** 231.03588	92 **U** 238.02891	93 **Np** [237.05]	94 **Pu** [244.06]	95 **Am** [243.06]	96 **Cm** [247.07]	97 **Bk** [247.07]	98 **Cf** [251.08]	99 **Es** [252.08]	100 **Fm** [257.10]	101 **Md** [258.10]	102 **No** [259.10]

[a]The labels on top (1A, 2A, etc.) are common American usage. The labels below these (1, 2, etc.) are those recommended by the International Union of Pure and Applied Chemistry.

The names and symbols for elements 110 and above have not yet been decided.

Atomic weights in brackets are the masses of the longest-lived or most important isotope of radioactive elements.

Further information is available at http://www.webelements.com

The production of element 116 was reported in May 1999 by scientists at Lawrence Berkeley National Laboratory.

Table of Contents

The Search for the Elements

> *"Mendeleev let the numbers (the atomic weights of the elements) speak. He did not let the numbers tyrannize him. He had the courage to leave blanks in his table, the elements that skillful hands had yet to discover."*
> *Roald Hoffmann and Vivian Torrence*

The history of the discovery of the elements is in large part the history of the development of methods that allowed scientists to identify elements and to separate them from more complex substances. Some metallic elements, such as gold and silver, can be found in their pure state in the earth. Other metallic elements—for example, iron—are found in nature as a metal ore (a compound made up of the metal and another element). Iron can be obtained in pure form by heating it with carbon in a process that chemists call *reduction.* Some elements are found in nature as gases. Even though gaseous elements, such as oxygen and nitrogen, are quite common because they are components of the air we breathe, they were not recognized as elements until relatively late in the history of the search for the elements.

Each element has its own interesting history of discovery. Often, the element's name reveals clues about its origin and early uses. Here are some examples:

Element	Origins of Its Name
Gold	Its name comes from the Latin *aurum*, meaning "shining dawn," a reference to its appearance.
Mercury	Also named because of its appearance. Known as *hydrargyrum* in Latin, meaning "liquid silver." Later named mercury after the planet.
Phosphorus	From the Greek, meaning "light bearer." One type of phosphorus will catch fire in air.
Nickel	German miners called the mineral "Kupfernickel," or "devil's copper."
Oxygen	From the Greek, meaning "acid former" and named by Antoine Lavoisier, who is often considered the father of modern chemistry.
Helium	From the Greek *helios,* meaning "the sun." Discovered by looking at characteristics of the light that comes from the sun.

One of the most inspiring stories in the history of chemistry is that of the Russian chemist Dmitri Mendeleev, who organized the experimental facts known about the elements in the 1860s into what is now the periodic table of the elements. On the basis of the earlier work of John Dalton, Mendeleev knew some important concepts about elements and the atoms that constitute them:

1. Elements are made up of extremely small, indivisible particles called atoms.
2. All atoms of an element have the same properties, including mass.*
3. Atoms can combine in simple whole-number ratios to form compounds.

Mendeleev came up with the idea for the periodic table while he was writing a book about the principles of chemistry. He wrote the properties of each element on a card, and when he looked at different arrangements of the cards, he noticed that the chemical properties of the elements were a repeating, or *periodic*, function of the weights of the atoms of each element. This observation became the foundation of the periodic table of the elements.

Self-Test 1

1. Mendeleev used the process of interpolation to predict the existence and properties of unknown elements. *Interpolation* is predicting something *within* the range of your data. By contrast, *extrapolation* is predicting something *beyond* the range of your data.

   Consider the following hypothetical table of average heights and weights of people:

Height (ft)	Weight (lb)
5.0	100
5.5	140
6.0	180

   a. What is the expected average weight for someone who is 5.8 feet tall? Is this an interpolation or an extrapolation of the data?

   b. What is the expected average weight for someone who is 6.5 feet tall? Is this an interpolation or an extrapolation of the data?

---

* We now know that atoms of an element can have slightly different masses. Atoms of an element that have different masses are called *isotopes*. See Unit 2, "Atoms and Subatomic Structure."

2. Mendeleev found that the physical properties of an element were related to the elements immediately above, below, and to both sides of it on the periodic table.

Relative Positions of S, Te, As, Br, and Se in Mendeleev's Table:

	S	
As	Se	Br
	Te	

Using the properties of S, Te, As, and Br from the table on page 10, calculate the expected atomic weight and density of selenium by interpolating. Compare your interpolated properties with the modern values.

Classifying Matter

Atomic theory establishes atoms as a basic unit of matter. How do atoms interact? How do they combine to make up the substances that we see and use every day? What is the difference between a single atom and a large number of atoms? To begin to answer these questions, chemists classify matter in several ways. One classification method is by physical state: *gas, liquid,* or *solid.* At the particulate level, the three common physical states can be distinguished by the interaction and motion of the particles. In a solid, the particles (atoms or molecules) are relatively fixed in space. In a liquid, the molecules are close together, but not fixed in space and are moving rather freely. In a gas, the particles have little interaction, are far apart, and move freely. Many substances occur as gas, liquid, or solid and can be transformed from one state to another by a change in temperature.

Matter can be classified as *homogeneous* and *heterogeneous.* Matter that is homogeneous is uniform throughout: No one portion of the material is different or distinguishable from any other portion. Materials that do not have this uniformity are heterogeneous. A piece of copper wire and a saltwater solution are homogeneous, while a sample of soil and a mixture of salt and pepper are heterogeneous. The salt and water in the saltwater solution can be separated by evaporating the water. When the components of the homogeneous matter can be separated by a physical process such as evaporation or distillation, the mixture is a *solution.* If the components can't be separated by physical means, then the material is a *pure substance.*

Self-Test 2

1. Place each of the following in the appropriate box:
   a. A mixture of sand and iron filings
   b. A mixture of water and alcohol
   c. Pure water
   d. Hydrogen gas

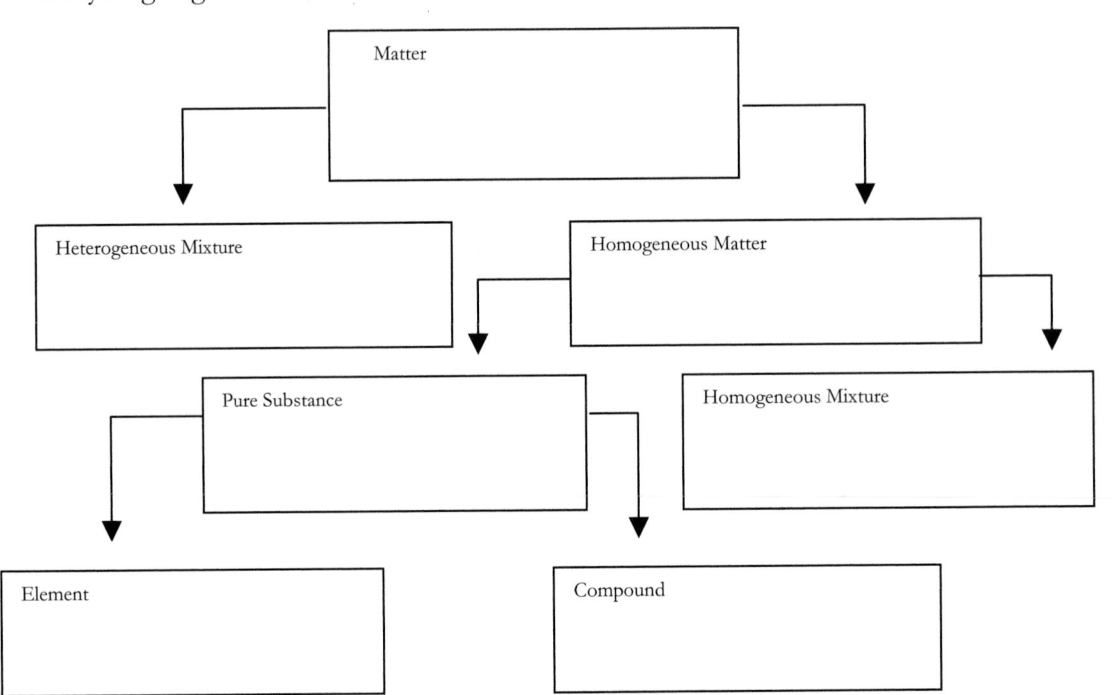

The International System of Units

In 1960, a group of scientists from many fields and many countries agreed upon a set of metric units that would serve as a standard for scientific communication. This standard set of units is known as the *International System of Units* and is abbreviated *SI.* (The abbreviation is derived from the French spelling *Système International d' Unités.*) Seven quantities are the foundation for SI, and each quantity has a *base unit* in which it is expressed. Table 1 lists the base units for length, mass, and time, along with their abbreviations and their relationships to common U.S. units.

The three SI base units listed in Table 1.1 were chosen because they correspond to magnitudes that are convenient for everyday measurement (i.e., at the macroscopic scale). However, chemists often work with tiny quantities, such as those used to express the diameter of a hydrogen atom, or huge quantities, such as the number of particles in a kilogram of carbon. These numbers are beyond the range of our senses and cannot be conveniently expressed in standard notation in SI units. Thus, the system of scientific notation is used to express very small and very large quantities.

Quantity	Base Unit	Abbreviation	U.S. Equivalent
Length	meter	m	39.37 inches
Mass	kilogram	kg	2.205 pounds
Time	second	s	1 second

Table 1.1. Three SI Base Units Commonly Used in General Chemistry.

Metric–U. S. conversions are made by applying dimensional analysis. For example, to convert a measured quantity of 4.00 inches to centimeters, we write

$$4.00 \text{ in.} \times \frac{2.54 \text{ cm}}{\text{in.}} = 10.2 \text{ cm}$$

Scientific notation is a method of expressing numbers as a product of two factors. The first factor is a number that is greater than or equal to 1, but less than 10. The second factor is 10 raised to a power. The power of 10, or *exponent,* is positive for numbers greater than 10 and negative for numbers less than 1. Table 1.2 gives examples showing both the ordinary decimal form and the exponential form for some quantities. Notice how scientific notation eliminates the need to write a long list of zeros in very small and very large numbers.

Quantity	Ordinary Decimal Form	Scientific Notation
Diameter of a hydrogen atom	0.000 000 000 074 m	7.4×10^{-11} m
Mass of a hydrogen atom	0.000 000 000 000 000 000 000 000 11 kg	1.1×10^{-25} kg
Number of molecules in 2.0 g of hydrogen	600 000 000 000 000 000 000 000	6.0×10^{23}

Table 1.2. Examples of Quantities Expressed in Scientific Notation.

Metric Prefixes To further simplify the expression of measured quantities, scientists use prefixes with metric units to represent powers of 10. Table 1.3 lists metric prefixes frequently used in chemistry. Notice that each prefix has an abbreviation and an equivalent power of 10. Your instructor may require you to memorize some or all of the prefixes and their associated powers of 10.

Prefix	Abbreviation	Power of Ten	Example
pico-	p	10^{-12}	picogram, pg
nano-	n	10^{-9}	nanometer, nm
micro-	μ	10^{-6}	microsecond, μs
milli-	m	10^{-3}	milliliter, mL
centi-	c	10^{-2}	centimeter, cm
kilo-	k	10^{3}	kilometer, km
mega-	M	10^{6}	megahertz, MHz
giga-	G	10^{9}	gigabyte, GB

Table 1.3. Common Metric Prefixes and Equivalent Powers of 10.

Self-Test 3

1. What is your height in inches? in meters? in nanometers?

2. Determine the appropriate SI units for each of the following:

   a. Density

   b. Area

   c. Velocity (speed)

   d. Force (mass times acceleration) (*Hint*: Acceleration is the change in velocity of an object with respect to time—a velocity change over a time interval)

   e. Pressure (force divided by area)

Mathematical Operations in Scientific Notation

Addition and Subtraction To add or subtract numbers that are expressed in scientific notation, use the "enter exponent" key on your calculator to express the power of 10. This key is usually labeled "EE" or "EXP". *Do not use the 10^x or y^x key for exponents in scientific notation:* The calculator usually will express the result in scientific notation automatically. For example,

$3.25 \times 10^3 + 4.66 \times 10^4 =$

Calculator key sequence: [3] [.] [2] [5] [EE] [3] [+] [4] [.] [6] [6] [EE] [4] [=]

Result: 4.985×10^4

Multiplication To multiply numbers that are expressed in scientific notation, use your calculator. For example,

$(3.1 \times 10^2) \times (5.2 \times 10^4) =$

Calculator key sequence: [3] [.] [1] [EE] [2] [×] [5] [.] [2] [EE] [4] [=]

Result: 1.612×10^7

Division To divide numbers that are expressed in scientific notation, again, let your calculator do the work. For example,

$$\frac{7.5 \times 10^2}{5.9 \times 10^{-4}} =$$

Calculator key sequence: [7] [.] [5] [EE] [2] [÷] [5] [.] [9] [EE] [4] [+/–] [=]

Result: $1.271... \times 10^6$

Powers To raise a number expressed in scientific notation to a power, use the y^x key on your calculator. For example,

$(2.5 \times 10^2)^3 =$

Calculator key sequence: [2] [.] [5] [EE] [2] [y^x] [3] [=]

Result: 1.5625×10^7

Measured Quantities and Significant Figures

There is a degree of uncertainty in every measurement, and this uncertainty, by convention, is reflected in the last recorded digit of any measured quantity. If several people measure the same distance with a ruler, their measurements will probably differ in the last digit. However, the measurements will cluster around the true value. Some will be equal to the true value, some will be higher, and some will be lower. The average of a series of measurements is generally considered to be the most accurate value.

The number of *significant figures* in a measurement is the total of the number of digits known with certainty, plus the one uncertain digit. The digits known with certainty are those which can be read precisely from the measuring instrument. The last digit recorded in any measured quantity—the uncertain digit—is estimated.

When a measurement is written in scientific notation, the first factor represents the significant digits in the measured quantity. The second factor, the power of 10, represents the magnitude of the measurement, or the number of decimal places, and therefore has nothing to do with the number of significant figures in the quantity. If the three-significant-figure measured quantity 134 pounds is written in scientific notation as 1.34×10^2 pounds, the number of significant figures cannot change, so this must also have three significant figures.

The following rules summarize the conventions used by chemists working with measured quantities:

1. The last digit expressed in a measured quantity is the uncertain, or estimated, digit.

2. Zeros that serve as placeholders in ordinary decimal notation are not significant. For example, the measured quantity 0.001 m has one significant figure. This fact becomes more apparent when we write the quantity in scientific notation: 1×10^{-3} m.

3. In adding and subtracting numbers representing measured quantities, the number of decimal places in the sum or difference is limited by the number of decimal places in the measured quantity that has the least number of decimal places. For example, 0.152 g + 0.26 g = 0.41 g.

4. When multiplying or dividing numbers representing measured quantities, the number of significant figures in the result is the number of significant figures in the factor with the least number of significant figures. For example, $(1.5 \times 10^2) \times (1.350 \times 10^3) = 2.0 \times 10^5$.

Self-Test 4

1. Express the following numbers in ordinary decimal form or scientific notation (the first line has been completed as an example).

Ordinary Decimal Form	Scientific Notation
0.683 kg	6.83×10^{-1} kg
1365 s	
	1.034×10^1 m
300 000 000 $m \cdot s^{-1}$	
	$(1.75 \times 10^5)\ (2.0 \times 10^5)$
$1605 + 3.22 \times 10^2$	
	$(112.33)^{1.5}$

2. How many significant figures would be appropriate for each of the quantities that follow? Explain each answer.

   a. Your height

   b. Your weight

   c. The speed of a car as read from a speedometer

3. Evaluate each of the following expressions, writing your answers in scientific notation and with the correct number of significant figures:

   a. $6.42 \times 10^4 + 3.5 \times 10^3 =$

   b. $\dfrac{2.00 \times 10^5}{4.0 \times 10^3} =$

   c. $(5 \times 10^2) \times (8.0 \times 10^3) =$

   d. $(4.00 \times 10^{-2})^3 =$

Workshop: The Search for the Elements

Consider the following list of some of the properties of the elements known to Mendeleev:

Element	Symbol	Atomic Weight (Relative to C at 12.01)	Density (g/cm^3)	Melting Point (K)	Common Compound with Hydrogen or Fluorine
Aluminum	Al	26.98	2.70	934	AlH_3
Antimony	Sb	121.76	6.69	904	SbH_3
Arsenic	As	74.92	5.73	1090	AsH_3
Beryllium	Be	9.01	1.85	400	BeH_2
Bromine	Br	79.90	3.12	266	HBr
Boron	B	10.81	2.34	2352	BH_3
Calcium	Ca	40.08	1.55	1112	CaH_2
Carbon	C	12.01	1.9–2.3 (graphite)	3820	CH_4
Chlorine	Cl	35.45	0.0032	172	HCl
Fluorine	F	19.00	0.0017	54	HF
Indium	In	114.82	7.31	430	InH_3
Lithium	Li	6.94	0.53	454	LiH
Magnesium	Mg	24.31	1.74	922	MgH_2
Nitrogen	N	14.01	0.0013	63	NH_3
Oxygen	O	16.00	0.0014	55	H_2O
Phosphorus	P	30.97	1.82 (white)	317 (white)	PH_3
Potassium	K	39.10	0.86	336	KH
Scandium	Sc	44.96	2.99	1814	ScH_3
Selenium	Se	78.96	4.79 (gray)	490 (gray)	H_2Se
Silicon	Si	28.09	2.33	1683	SiH_4
Sodium	Na	22.99	0.97	371	NaH
Sulfur	S	32.07	2.07 (rhombic)	386 (rhombic)	H_2S
Tellurium	Te	127.60	6.24	723	H_2Te
Tin	Sn	118.71	5.75 (gray)	505 (gray)	SnH_2
Zinc	Zn	65.39	7.13	693	ZnF_2

Generally, Mendeleev arranged the elements in order of increasing atomic weight. Let's take a look at some of the elements in order to understand his approach.

1. Obtain a pad of self-stick notes, and write the symbol, atomic weight, density, melting point, and common compound with hydrogen or fluorine of each element on individual notes, from the information given in the preceding table. This task will go quickly if you divide it among the members of your group.

   Next, reproduce the table that follows on the whiteboard or chalkboard. Insert the self-stick notes representing the elements in order of increasing atomic weight, from left to right, in the blank spaces. Some symbols are already filled in. Question marks are in two of the places that Mendeleev left empty in his table. Skip these places when your group fills in the table. Omit the boxes that are darkened.

   Leave the periodic table that you've constructed on the board, and use it as necessary as you consider the remaining questions in this unit.

H																	He
																	Ne
																	Ar
			Ti	V	Cr	Mn	Fe	Co	Ni	Cu		?	?				Kr
Rb	Sr	Y	Zr	Nb	Mo	Tc	Ru	Rh	Pd	Ag	Cd					I	Xe

What are the differences between your periodic table, based on increasing atomic weights, and the modern periodic table? Why do these differences exist?

2. *We suggest that you consider the round-robin approach for this question.*

While constructing his periodic table, Mendeleev considered a property known as "valence" or "combining power." As examples, hydrogen has a combining power of 1: It cannot combine with more than one atom at a time. Nitrogen has a combining power of 3, because it can combine with three hydrogens to form the compound NH_3. Oxygen has a typical combining power of 2. Fluorine combines with 1 hydrogen. List the combining powers of the elements from lithium to chlorine in the next table. We have already filled in the valence for lithium as an example. Use the data in the table on page 10 to complete the following table:

Element	Li	Be	B	C	N	O	F
Valence	1						
Element	Na	Mg	Al	Si	P	S	Cl
Valence							

Describe the pattern that you see. What is the relationship between the position of an element in the periodic table and its valence?

Questions 3–4: Paired problem solving works well for these questions.

3. Mendeleev called the missing elements in the table for Question 1 eka-aluminum (now called gallium) and eka-silicon (now called germanium). Not only did he predict the existence of then-unknown elements, but also, by the process of interpolation, he estimated the chemical and physical properties of the unknown elements. Based on the pattern in the combining powers of the elements, predict the following:

   a. A common compound that will form from germanium and hydrogen.

   b. A common compound that will form from gallium and fluorine.

Question 4: Break into pairs to answer this question. Your Workshop leader will assign one or two parts of the question to each pair. When you have completed your answers, present them to the group. For this question, consider the following list:

A. water
B. iron
C. a cup containing grains of sand and particles of iron
D. a cup containing crystals of sugar and particles of glass
E. iron oxide
F. sand
G. air
H. sugar dissolved in water
I. gold
J. nitrogen
K. oxygen
L. carbon dioxide

4. a. Speculate on what could have been the first element discovered by early humans (before modern science). Explain how you arrived at your answer.

b. Which compound can be reduced to the pure element by reaction with carbon? Explain.

c. How could you separate the sand and iron?

d. How could you separate the sugar and glass?

5. In ancient times, natural philosophers proposed a classification system that had four elements: earth, air, water, and fire. How would these substances be classified today?

6. The speed of light is 299,792,458 meters per second.

   a. Express the speed of light in exponential form.

   b. How many significant figures does this number have?

   c. Express the speed of light in kilometers per second.

   d. Express the speed of light in miles per hour.

   e. A light-year is the distance that light travels in 1 year. Calculate this distance in miles. Express your answer in exponential form and with the correct number of significant figures.

Consider the figures shown next for Questions 7–10. In each case, the double-arrow line represents a distance of 1.00 inch. We suggest the round-robin approach to these problems.

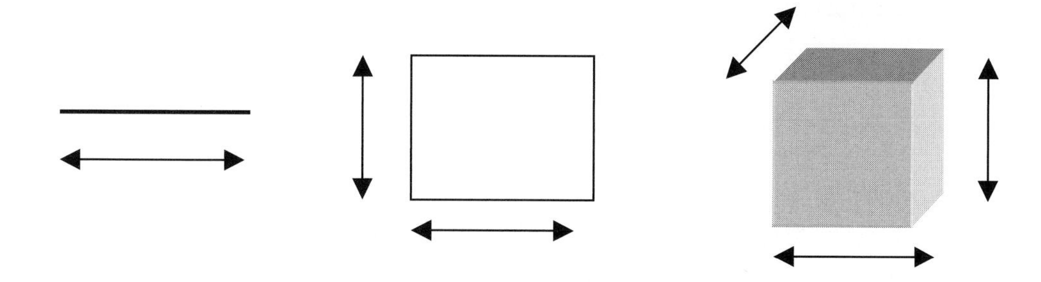

7. Using the square and the cube, convert the square inch and the cubic inch to square centimeters and cubic centimeters.

8. Convert 1.2 in^2 to cm^2.

9. Convert 5.881 in^3 to cm^3.

10. Since the meter is the SI unit of length, the cubic meter is the preferred unit of volume. Chemists never express volume in cubic meters, however. The liter and the milliliter are the most common units of volume. How many cubic meters are in a liter? How many milliliters are in a cubic meter?

Atoms and Subatomic Structure

> *"No new creation or destruction of matter is within the reach of chemical agency. We might as well attempt to introduce a new planet into the solar system, or to annihilate one already in existence, as to create or destroy a particle of hydrogen."*
> *John Dalton*

The understanding that matter is composed of different elements evolved over thousands of years. Substances such as gold and silver were known in ancient times, but they were not understood to be elements. The alchemists, who did not understand elements or atoms, tried to change substances into gold because it was such a valuable substance. They never succeeded because, as we know now, matter is made up of atoms that cannot be changed by ordinary chemical methods. By the end of the 19th century, chemists had formulated an atomic theory of matter which stated that the atom was the smallest particle of an element. This atomic theory also recognized that atoms were composed of smaller, subatomic particles. Today we recognize more than 110 different elements.

The Chemists' Collection of Subatomic Particles

Particle beam technology became available by the turn of the century, and it enabled scientists to discover the three subatomic particles listed in the following table:

Particle	Mass (g)	Charge
Proton	1.672×10^{-24}	+1
Neutron	1.675×10^{-24}	0
Electron	9.109×10^{-28}	–1

The nature of the electron was determined by experiments on the interaction between beams of electrons (cathode rays) and magnetic fields. In addition, as a result of Rutherford's alpha-particle experiments in which helium nuclei were projected through metal foils, a new picture of the atom emerged. In this model, the atom is viewed as mostly empty space, with the mass concentrated in a dense, compact core of protons and neutrons known as the nucleus. The space surrounding the nucleus contains the atom's electrons.

Symbols for the Elements

The periodic table of the elements shows the complete collection of known elements, with each elemental name represented by a one- or two-letter symbol. (Three-letter symbols are used temporarily until the element is given a final name.) These symbols are sometimes written with a superscript or a subscript (or both) preceding them. In this case, the symbols are called *nuclear symbols*, and they are used to distinguish among the isotopes of an element. All atoms of an element have the same number of protons, but the number of neutrons in an atom can vary. The superscript in a nuclear symbol represents the total number of particles in the nucleus: the sum of the number of protons and the number of neutrons. The subscript indicates the number of protons. Since an atom is electrically neutral, the number of electrons can also be determined from the nuclear symbol, because it is equal to the number of protons.

The nuclear symbol of any element is given by

$$^{A}_{Z}\mathrm{Sy}$$

where
A = mass number = number of protons + number of neutrons,
Z = atomic number = number of protons,
Sy = elemental symbol.

For example, the symbol for carbon-12 is $^{12}_{6}\mathrm{C}$. The symbol tells us that an atom of carbon-12 has 6 protons (subscript) and 12 protons + neutrons (superscript). The difference, 12 – 6 = 6, is the number of neutrons. We also know that, since the atom is electrically neutral, the carbon atom has six electrons.

In ordinary chemical reactions, atoms never lose or gain protons or neutrons. However, they do lose or gain electrons. In fact, the chemical properties of an element are largely determined by the arrangement of the outer electrons of its atoms.

Mass Units for Elements

Chemists use a variety of units to express both the absolute and the relative mass of each of the elements. These units and the conventions governing their use are the subject of this section.

Atomic Mass Units (amu) Chemists have chosen carbon-12 as the standard for atomic masses. The carbon-12 atom is assigned a mass of exactly 12 atomic mass units. This means that

$$1 \text{ amu} = \frac{1}{12} \text{ the mass of one carbon-12 atom}$$

Also, since 6.02×10^{23} carbon-12 atoms have a mass of 12 grams,

$$1 \text{ amu} = \frac{1}{6.02 \times 10^{23}} \text{ grams} = 1.66 \times 10^{-24} \text{ grams}$$

Isotopes of an element necessarily have different masses because they have different numbers of neutrons. For example, carbon-13 ($^{13}_{6}C$), which has six protons, six electrons, and seven neutrons, has a mass of 13.00335 amu.

Carbon-12 and carbon-13 atoms are both present in any ordinary sample of carbon. The fractional abundance of carbon-12 is 0.9890, and that of carbon-13 is 0.0110. The fractional abundances for these isotopes must add up to 1. Fractional abundance can also be expressed as a percentage, and the numbers are 98.90% and 1.10% for carbon-12 and carbon-13, respectively. These usually add to 100%, although experimental uncertainty may cause slight deviations.

Atomic Mass (Weight) The atomic mass (weight) of an element is defined as the average mass of the naturally occurring isotopes of the elements, weighted to account for their percentage abundance. The unit of atomic mass is the amu.

Self-Test 1

1. Fill in the blanks in the chart that follows. Be sure to include ionic charges when appropriate. The first row is completed as an example.

Nuclear Symbol	Number of Protons	Number of Neutrons	Number of Electrons	Atomic Number (Z)	Mass Number (A)
$^{12}_{6}C$	6	6	6	6	12
$^{14}_{7}N$			7		
	7	8	7		
			18	20	40
$^{17}O^{2-}$				8	
^{56}Fe			26		
$^{19}F^{-}$				9	

2. Refer to the chart you completed in Question 1, and identify

   a. All of the isotopes.

   b. The atoms and ions that are isoelectronic (those with the same number of electrons).

3. Which element is the standard for assigning the atomic mass of all elements? Explain.

4. Arrange the following in order of increasing mass: (a) the mass of a proton, (b) the mass of a grain of sand, (c) 15.999 amu, (d) the mass of a single carbon atom, and (e) the mass of an electron.

The Mole A mole (mol) is 6.02×10^{23} atoms, molecules, or particles of any kind. The number 6.02×10^{23} is called *Avogadro's number,* and it is given the symbol N_A.

Molar Mass The molar mass of an element is the mass in grams of 1 mole of atoms of that element. Therefore, molar mass units are grams per mole (g/mol). The molar mass of an element is numerically equal to the atomic mass of the element. For example, the atomic mass of carbon is 12.011 amu and its molar mass is 12.011 g/mol.

Self-Test 2

1. Coin A has a mass of 10.00 g. Coin B has a mass of 12.00 g. If you have 25 coins of type A and 75 of type B, what is the average mass for this collection of coins?

2. Using your solution to Question 1 as an example, write a general formula that can be used to calculate the average mass for any number of coins A and B.

3. Carbon-12 has a mass of 12.00000 amu and Carbon-13 a mass of 13.00335 amu. The fractional abundance of C-12 is 98.90% and the fractional abundance of C-13 is 1.10%. Use this information and your formula from Question 2 to calculate the atomic weight of carbon.

Compounds: Atoms in Combination

Atoms of two or more elements can combine to form new substances called *compounds. Ionic compounds* form when one atom transfers one or more electrons to another atom. *Covalent* or *molecular compounds* form when atoms share one or more pairs of electrons.

Ions Some elements have a tendency to gain electrons and therefore become negatively charged ions, which are called *anions.* Other elements have a tendency to lose electrons and become positively charged ions, known as *cations.* Examples include a chlorine atom gaining an electron to become a chloride ion (an anion), symbolized

$$Cl + e^- \rightarrow Cl^-$$

and a sodium atom losing an electron to become a sodium ion (a cation), symbolized

$$Na \rightarrow Na^+ + e^-$$

Ionic Compounds Cations and anions combine to form ionic compounds. The force of attraction between the positive and negative ionic charges holds these compounds together. Certain groups of atoms, such as SO_4^{2-} or NH_4^+, commonly occur together as anions or cations. These polyatomic ions can also form ionic compounds.

Ionic compounds exist as extended arrays of anions and cations. The chemical formula of an ionic compound gives the smallest whole-number ratio of cations to anions. For example, the formula for sodium chloride, NaCl, tells us that the ratio of sodium ions to chloride ions is one to one. However, a macroscopic sample of a sodium chloride (table salt) crystal contains billions and billions of sodium ions and chloride ions.

Molecules Compounds such as water (H_2O), ammonia (NH_3), and buckminsterfullerene (C_{60}) are molecules. They are molecular rather than ionic because the atoms in the molecules are held together by sharing electrons. Many compounds and some elements exist as molecular substances.

Examples of molecular substances include the well-known water molecule, which has two hydrogen atoms, each sharing a pair of electrons with an oxygen atom:

The ammonia molecule looks like a pyramid in which three hydrogen atoms are covalently bonded to a single nitrogen atom:

The buckminsterfullerene molecule exists as a discrete unit in which 60 carbon atoms are covalently bonded to each other in a shape that resembles a soccer ball:

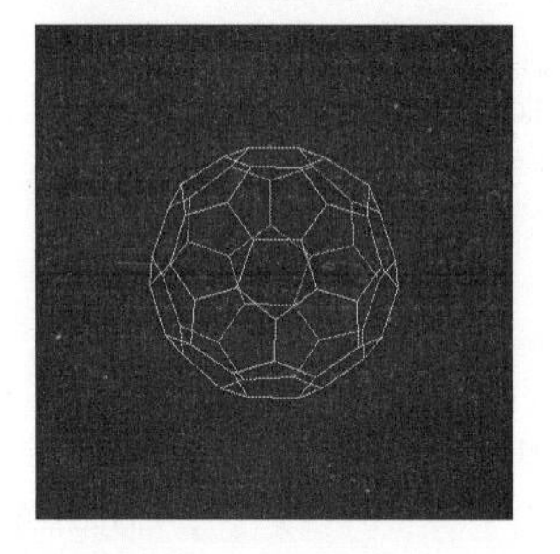

We must caution you that the distinctions between ionic and molecular compounds are not a simple case of either one or the other, but instead represent two extremes. The definitions given are only a first step in understanding how atoms join together in the process of chemical bonding.

Self-Test 3

1. Write the formula of each of the following: potassium ion, cobalt(III) ion, bromide ion, phosphate ion.

2. Write the name of each of the following: Ca^{2+}, Ag^{+}, S^{2-}, ClO_3^{-}.

Workshop: Atoms and Subatomic Structure

Questions 1–8: We suggest paired problem solving followed by group discussion for these questions.

1. Naturally occurring boron, atomic number 5, is a mixture of two isotopes: boron-10 and boron-11. Work in pairs to sketch representations of each of these atoms, specifically illustrating the difference between the isotopes. When each pair is finished, compare sketches and discuss and resolve any differences.

2. Boron is present in small amounts in the earth. It is a necessary nutrient for plants, although it is toxic to humans in large amounts. The boron-10 isotope has a mass of 10.0129 amu and boron-11 has a mass of 11.00931 amu. The atomic mass of a natural sample of boron containing both isotopes of boron is 10.811 amu.

   a. Without using a calculator, choose the best estimate among the following for the percentage abundance of the two boron isotopes:

      (i) 40% ^{10}B and 60% ^{11}B
      (ii) 80% ^{10}B and 20% ^{11}B
      (iii) 20% ^{10}B and 80% ^{11}B
      (iv) 60% ^{10}B and 40% ^{11}B

   b. Calculate the actual percentage abundance of the boron isotopes. How does your calculated value compare with your estimate from part (a)?

3. A soil sample is found to contain 9.00×10^{-6} g of boron. Express this mass in amu.

4. Nitrogen is 99.63% ^{14}N at 14.00307 amu and 0.37% ^{15}N at 15.00011 amu. Determine the atomic mass of nitrogen. How does your calculated value compare with the value given in the periodic table?

5. Why is the identity of an element determined by the number of protons in its nucleus? Why can't an element's identity be determined by the number of neutrons or the number of electrons?

6. Assume that the best analytical balance available is accurate to within 0.00001 g. What is the smallest number of carbon atoms that can be detected by this balance? How many moles is that number of carbon atoms? Which method of expressing quantity—number of atoms or moles of atoms—is more convenient?

7. Solve each of the following:

a. The atomic mass of carbon is 12.011 amu. What is the mass, in grams, of 6.02×10^{23} carbon atoms?

b. What is the mass of 1.000 mol of sodium atoms?

c. How many neon atoms are in a 1.0-kg sample of the pure element?

8. The density of diamond, a form of pure carbon, is 3.5 g/cm^3. The radius of a carbon atom is 77 picometers, and the volume of a sphere is given by $V = (4/3)\pi r^3$.

a. What is the volume of a carbon atom?

b. What is the density of a carbon atom, in g/cm^3?

c. Compare the density of a carbon atom with the density of diamond. Why is there a difference in densities?

d. The nucleus of a carbon atom has a radius of 1.5×10^{-5} pm. What is the volume of the nucleus?

e. What is the density of the nucleus of a carbon atom, in g/cm^3?

f. What do the preceding calculations tell you about the Rutherford alpha-particle experiments?

9. The C_{60} molecule has probably been a product of campfire burning since the earliest history of human civilization, yet it was not formally proposed to exist until 1985, and its existence was finally verified in 1990. The 1996 Nobel Prize in Chemistry was awarded to R. F. Curl, Jr., H. W. Kroto, and R. E. Smalley for their discovery of this general class of molecule, known as fullerenes. As indicated by the formula, the molecule consists of 60 carbon atoms. Because of its shape, it was named buckminsterfullerene in honor of Buckminster Fuller, a 20th-century visionary engineer, architect, scientist, and philosopher.

a. What is the mass of the C_{60} molecule, in amu?

b. What is the molar mass of C_{60}?

c. Is every molecule of C_{60} identical? Explain.

10. *We suggest a round-robin approach, completing the table on the whiteboard or chalkboard.* Complete the table that follows by writing both the formula and the name of the compound formed by the combination of the cation in the top row with the anion in the left column. The first box is completed as an example.

	Na^+	Hg_2^{2+}	Fe^{3+}	Cu^{2+}	Cr^{3+}
Cl^-	NaCl sodium chloride				
NO_3^-					
SO_4^{2-}					
N^{3-}					
CO_3^{2-}					

Unit

3

Introduction to Stoichiometry

Stoichiometry comes from the Greek words stoicheion, *meaning element or part, and* metron, *meaning measure. Stoichiometry counts individual particles by making macroscopic measurements.*

Many chemical reactions have been known and applied for thousands of years. For example, when iron oxides are heated in the presence of carbon, pure iron is obtained. Yet this practical knowledge revealed nothing about the elemental nature of matter. It took careful measurements of the masses of substances consumed and produced in chemical reactions to reveal the nature of elements and how they combined with each other to form compounds. By the 19th century, stoichiometry had led to a deeper understanding of the atomic and molecular basis of chemistry, which had previously eluded the human mind. Consequently, stoichiometry played a pivotal role in the history of chemistry and continues to be the starting point for a detailed investigation into modern chemistry.

Molar Masses from Compound Formulas

The formulas of ionic compounds give the ratio of ions in the crystal lattice. The formula NaCl indicates a one-to-one ratio of Na^+ ions to Cl^- ions in the NaCl lattice. In contrast, a molecular formula gives the actual description of the atoms in each individual molecule. For example, the molecular formula of hydrogen peroxide is H_2O_2, and each hydrogen peroxide molecule has the structure H—O—O—H. *If* hydrogen peroxide were an ionic compound (it's not), its formula would be written HO, reflecting the one-to-one ratio of hydrogen atoms to oxygen atoms in the compound. Chemical formulas that give the smallest whole-number ratios of atoms in a molecule are called *empirical formulas.*

The molar mass of NaCl is calculated as the sum of the molar masses of its ions:

$$22.99 \text{ g/mol} + 35.45 \text{ g/mol} = 58.44 \text{ g/mol}$$

The molar mass of H_2O_2 is calculated as the sum of the molar masses of the atoms that make up the molecule:

$$2 \times 1.008 \text{ g/mol} + 2 \times 16.00 \text{ g/mol} = 34.02 \text{ g/mol}$$

Percentage Composition

The percentage composition of a compound is the percentage by mass of each element in the compound. To calculate a percentage, we use the formula

$$\%\text{ of X} = \frac{\text{parts of X}}{\text{total parts}} \times 100$$

Thus, for hydrogen peroxide,

$$\%\text{H} = \frac{2 \times 1.008 \text{ g/mol}}{34.02 \text{ g/mol}} \times 100 = 5.926\%$$

$$\%\text{O} = \frac{2 \times 16.00 \text{ g/mol}}{34.02 \text{ g/mol}} \times 100 = 94.06\%$$

The sum of the percentages must be 100, so we can use it as a check:

$$5.926\% + 94.06\% = 99.99\%$$

It is acceptable to have the sum be off by ±0.02% or so, because of rounding in each of the individual percentage calculations.

Self-Test 1

1. Fill in the following table, using a periodic table, your lecture notes, and Unit 2 for additional information.

Compound	Empirical Formula	Molar Mass	Moles in 100.0 g
C_2H_6	CH_3	30.07 g/mol	3.33 mol
C_{60}			
C_2H_5OH			
$BaSO_4$			
$C_6H_{12}O_6$			

2. Calculate the percentage composition by mass of each of the following compounds:

Compound	Percentage Composition by Mass
C_2H_6	
C_{60}	
C_2H_5OH	
$BaSO_4$	
$C_6H_{12}O_6$	

Chemical Reactions

Chemists use equations to represent chemical reactions. The formation of water from the elements hydrogen and oxygen is represented by the equation

$$2\ H_2(g)\ +\ O_2(g)\ \rightarrow\ 2\ H_2O(\ell)$$

For this equation, and for any other chemical equation,

1. The reactants—hydrogen and oxygen—are written on the left side of the reaction arrow. The product—water—is written on the right side.

2. The reaction coefficients can be interpreted as the number of molecules or moles of reactants and products. The foregoing equation says that two molecules (or two moles) of hydrogen react with one oxygen molecule (or one mole of oxygen molecules) to form two water molecules (or two moles of water molecules).

3. State symbols are used to indicate whether the reactants and products are in the solid (s), liquid (ℓ), gas (g), or aqueous (aq) phase. Aqueous means that the substance is dissolved in water. In the preceding example, hydrogen and oxygen are in the gaseous state, and water is in the liquid state.

4. There is the same number of each kind of atom on each side of the equation. This "atom balance" is also called a "mass balance," because, if the atoms are balanced, their masses are also balanced.

Item 2 in the preceding list gives us a method for calculating the relative mole quantities required for a chemical change. For example, in the formation of water, there must always be twice the number of moles of hydrogen reacting as the number of moles of oxygen. So if 0.10 mol of $O_2(g)$ reacts, 0.20 mole of $H_2(g)$ must also react:

$$0.10 \text{ mol } O_2 \times \frac{2 \text{ mol } H_2}{1 \text{ mol } O_2} = 0.20 \text{ mol } H_2$$

The coefficients in the balanced chemical equation will always provide you with the mole ratio for any pair of species, either reactant or product.

Self-Test 2

1. For each of the following questions, first balance the skeleton (unbalanced) equation and then determine the answer:

   a. ____ $C_2H_5OH(\ell)$ + ____ $O_2(g)$ → ____ $CO_2(g)$ + ____ $H_2O(g)$
   How many moles of ethanol will react with 2.500 moles of oxygen gas?

   b. ____ $Fe(s)$ + ____ $O_2(g)$ → ____ $Fe_2O_3(s)$
   How many moles of iron will react with 3.00 moles of oxygen gas?

Determination of Molecular Formulas

Knowledge of the percentage composition and molar mass of a compound allows for the determination of its molecular formula. For example, benzene is 92.25% carbon and 7.74% hydrogen, and its molar mass is 78.11 g/mol. From these data, we can find the formula of benzene.

We begin by assuming that we have a 100-g sample of the compound. (The formula is the same no matter the size of the sample, so we can choose a convenient mass.) This translates to 92.25 g C and 7.74 g H.

Next, the number of moles of each element is determined:

$$92.25 \text{ g C} \times \frac{1 \text{ mol C}}{12.01 \text{ g C}} = 7.681 \text{ mol C}$$

$$7.74 \text{ g H} \times \frac{1 \text{ mol H}}{1.008 \text{ g H}} = 7.68 \text{ mol H}$$

The molecular formula of a compound tells you the ratio of the numbers of atoms of each element in the compound, which is also the ratio of moles of atoms of each element. In this case, by inspection, it can be seen that the C:H mole ratio is 1:1. This simplest formula ratio gives what is called the empirical formula, CH.

The molecular formula must have a 1:1 C:H mole ratio. It may be CH, C_2H_2, C_3H_3, and so on. We find the actual formula by calculating the ratio of the known true molar mass to the molar mass of the empirical formula, which we just determined as CH:

$$\frac{78.11 \text{ g/mol actual formula}}{(12.01 + 1.008) \text{ g/mol empirical formula}} = 6$$

There are 6 CH units in the actual molecular formula, so it is C_6H_6. Workshop questions 5 and 6 will give your team the opportunity to practice determining molecular formulas.

Self-Test 3

1. Acetic acid is 39.9% C, 6.7% H, and 53.4% O, and its molar mass is 60.0 g/mol. From these data, determine the molecular formula of acetic acid.

Workshop: Introduction to Stoichiometry

Use the group round-robin method, detailed in the following steps, to solve Questions 1 and 2:

STEP 1: The workshop leader will assign each person in the group a number starting from 1 and going up to the number of people in the group.

STEP 2: Person number 1 starts the problem by completing the first step in the solution. When the group agrees that the first step is correct, person number 2 completes the second step.

STEP 3: When the group agrees that the second step is correct, person number 3 does the third step.

STEP 4: Continue in this fashion until the solution to the problem is complete and all group members agree that the answer is correct. The person with the lowest number who has not yet contributed to the group begins the next problem and completes the first step.

STEP 5: Continue this pattern for the remaining round-robin questions.

1. The purpose of this question is to develop a conceptual understanding of the magnitude of Avogadro's number. Consider what would happen if 1 mole of pennies were distributed equally among the earth's population, which is currently estimated at 6×10^9.*

   a. How many pennies would each person get?

   b. If you spend a mole of pennies at the rate of 1 million dollars per day, how many years will it take to spend all of the pennies? Use dimensional analysis (factor–label) to construct a clear, easy-to-follow solution.

   c. Which has more monetary value, 15.00 moles of pennies or 0.800 moles of quarters?

   d. The mass of one penny is 2.45 g. What is the mass of 1 mole of pennies?

*This problem is adapted from a problem that appeared in American Chemical Society (1988), *ChemCom: Chemistry in the Community* (Dubuque, IA: Kendall/Hunt).

2. Consider the decomposition of hydrogen peroxide:

$$2\ H_2O_2(aq) \rightarrow 2\ H_2O(\ell) + O_2(g)$$

Build models of hydrogen peroxide molecules with pennies representing oxygen atoms and dimes representing hydrogen atoms.

a. Use your models to show the products of the reaction of 2 H_2O_2 molecules.

b. Use your models to show the products of the reaction of 4 H_2O_2 molecules.

c. How many of each of the product molecules will be formed from the reaction of 2 dozen molecules of H_2O_2? 4 dozen molecules of H_2O_2?

d. How many of each of the product molecules will be formed from the reaction of 2 million molecules of H_2O_2? 4 million molecules of H_2O_2?

e. How many of each of the product molecules will be formed from the reaction of 2 moles of H_2O_2 molecules? 4 moles of H_2O_2 molecules?

f. What is the relationship among your answers to the preceding questions? Explain why this relationship holds.

Questions 3 and 4: Paired problem solving is an effective way to deepen one's understanding of a problem or phenomenon. One person assumes the role of explaining while the other acts as a recorder, writing everything that the explainer says. After the problem is solved, review the transcript to discover the path that was taken. Change roles after Question 3. Record who plays each role in the activity.

3. How many moles of aluminum are in 1.35 grams? How many atoms?

4. What is the mass of 1.204×10^{24} copper atoms?

For Question 5, use the group round-robin approach described at the beginning of this workshop. (You may wish to divide your group into pairs and assign one element to each pair for parts a and b.)

5. Pyrophosphoric acid is composed of 2.27% hydrogen and 34.80% phosphorus. The remainder is oxygen.

   a. What is the mass of each element present in a 100.0-g sample?

   b. Determine the number of moles of each element in the 100.0-g sample.

   c. Find the simplest whole-number ratio of the number of moles of each element. This ratio can be found by dividing the number of moles of each element by the number of moles of the element with the smallest number of moles. Your result gives the empirical formula of pyrophosphoric acid.

   d. Calculate the molar mass of the pyrophosphoric acid empirical formula unit.

   e. The molar mass of pyrophosphoric acid is 177.97 g/mol. What is its molecular formula?

6. Vitamin C is an antioxidant that can be represented by the formula $C_xH_yO_z$, where x, y, and z are integers. Antioxidants are important in biochemistry partly because they remove potentially harmful oxidizing substances from the body. The purpose of this question is to determine the molecular formula of vitamin C by calculating the values of x, y, and z from the following data:

A 35.5-mg sample of vitamin C reacted with oxygen in a combustion apparatus, and 53.3 mg CO_2 and 14.4 mg H_2O were recovered.

Use a group brainstorming approach to solve parts a through d.

a. Write an unbalanced chemical equation for the combustion of vitamin C, identifying the reactants and products. The equation cannot yet be balanced, because vitamin C can be represented only as $C_xH_yO_z$, where x, y, and z are integers to be determined in a later step in the solution of the problem.

b. What is the reactant source of the carbon in the carbon dioxide product?

c. What is the reactant source of the hydrogen in the water product?

d. What is the reactant source of the oxygen in the carbon dioxide and water products?

Use a group round-robin approach to solve the remaining parts of the question.

e. How many moles of carbon were in the original vitamin C sample?

f. How many moles of hydrogen were in the sample?

g. How many moles of oxygen were in the sample?

h. What is the mole ratio of carbon to hydrogen to oxygen in vitamin C?

i. What is the empirical formula for vitamin C?

j. The molar mass of vitamin C is 176 g/mol. What is the molecular formula of vitamin C? In other words, what are the values of x, y, and z in $C_xH_yO_z$?

Strategies for Stoichiometry

> *"Solving problems is a practical art, like swimming, or skiing or playing the piano . . .*
> *and if you wish to become a problem solver you have to solve problems."*
> *George Polya*

Butane is an excellent fuel because it burns in air to give carbon dioxide, water vapor, and heat according to the following equation:

$$2\ C_4H_{10}(g)\ +\ 13\ O_2(g)\ \rightarrow\ 8\ CO_2(g)\ +\ 10\ H_2O(g) + \text{heat}$$

The balanced equation gives us all the information we need to determine the mass relationships among the reactants and products because the coefficients in the balanced equation can be interpreted as moles of substance. In the butane reaction, the coefficients tell us that 2 moles of butane react with 13 moles of oxygen to form 8 moles of carbon dioxide and 10 moles of water.

Suppose that the ordinary butane pocket lighter contains 2.325 grams of butane. How many grams of oxygen are needed to react with this amount of butane? How many grams of carbon dioxide would be formed when all the butane reacts with oxygen? These are questions that chemists typically ask about mass relationships in a chemical reaction. To answer those questions, chemists use a three-step dimensional analysis process that first converts the given mass to moles and then converts the moles of the given compound to moles of the unknown compound, using the coefficients in the balanced equation. In the third step, the moles of the unknown compound are converted to grams and the question is answered.

Convert grams of butane to moles:

$$2.325\ \text{g}\ C_4H_{10} \times \frac{1\ \text{mol}\ C_4H_{10}}{58.12\ \text{g}\ C_4H_{10}} = 0.04000\ \text{mol}\ C_4H_{10}$$

Convert moles of butane to moles of oxygen:

$$0.04000\ \text{mol}\ C_4H_{10} \times \frac{13\ \text{mol}\ O_2}{2\ \text{mol}\ C_4H_{10}} = 0.2600\ \text{mol}\ O_2$$

Convert moles of oxygen to grams:

$$0.2600 \text{ mol } O_2 \times \frac{32.00 \text{ g } O_2}{1 \text{ mol } O_2} = 8.320 \text{ g } O_2$$

These steps are summarized in the single equation that follows. Look at it carefully to see how all the factors work to move from grams of butane to grams of oxygen:

$$2.325 \text{ g } C_4H_{10} \times \frac{1 \text{ mol } C_4H_{10}}{58.12 \text{ g } C_4H_{10}} \times \frac{13 \text{ mol } O_2}{2 \text{ mol } C_4H_{10}} \times \frac{32.00 \text{ g } O_2}{1 \text{ mol } O_2} = 8.320 \text{ g } O_2$$

Self-Test 1

1. Use the balanced equation for the combustion of butane and the procedure in the preceding discussion to determine the mass of carbon dioxide that forms when 2.325 grams of butane react completely with oxygen.

2. A total of 20.00 g of zinc reacts with excess oxygen to form zinc oxide. Calculate the mass of oxygen needed to react with all of the zinc.

3. Lead reacts with lead(IV) oxide and sulfuric acid in a car battery according to the following equation:

$$Pb(s) + PbO_2(s) + 2\,H_2SO_4(aq) \rightarrow 2\,PbSO_4(s) + 2\,H_2O(\ell)$$

How many grams of $PbSO_4$ are formed when 103.6 g of lead reacts with PbO_2 and H_2SO_4? Assume that there are adequate quantities of PbO_2 and H_2SO_4 to use up all the lead.

In most chemical reactions, reactants and products are not present in the precise amounts required by the chemical equation. We will examine two common cases, one in which the reactant is present in excess and the other in which less product is produced than is stoichiometrically expected.

Reactants: One Reactant in Excess

In real reactions between butane and oxygen, there are almost never exactly 2 moles of butane and 13 moles of oxygen. The most common situation for combustion is where the reaction takes place in air, in which case an essentially infinite amount of oxygen is available to react. We call oxygen the *excess reactant,* because it is available in excess of the amount needed to react with all of the butane. Butane is called the *limiting reactant,* because it limits the amounts of products that can be formed—it will run out first. The initial quantity of limiting reactant will determine the amounts of products that can be formed.

Let's consider the burning of charcoal briquettes as an example. The primary reaction that occurs is

$$C(s) + O_2(g) \rightarrow CO_2(g)$$

If we were to begin with 18.0 g C and 80.0 g O_2, we would have

$$18.0 \text{ g C} \times \frac{1 \text{ mol C}}{12.01 \text{ g C}} = 1.50 \text{ mol C}$$

$$80.0 \text{ g } O_2 \times \frac{1 \text{ mol } O_2}{32.00 \text{ g } O_2} = 2.50 \text{ mol } O_2$$

According to the balanced chemical equation, 1 mole of product CO_2 forms from 1 mole of each of the reactants C and O_2.

If carbon is the limiting reactant, 1.50 mol of carbon dioxide will form:

$$1.50 \text{ mol C} \times \frac{1 \text{ mol } CO_2}{1 \text{ mol C}} = 1.50 \text{ mol } CO_2$$

If oxygen is the limiting reactant, 2.50 mol of carbon dioxide will form:

$$2.50 \text{ mol } O_2 \times \frac{1 \text{ mol } CO_2}{1 \text{ mol } O_2} = 2.50 \text{ mol } CO_2$$

The limiting reactant is the reactant that limits the amount of product that is formed. It produced the lesser amount of product. In this case, the limiting reactant is carbon, and 1.50 moles of carbon dioxide will form. Let's change this amount to grams:

$$1.50 \text{ mol } CO_2 \times \frac{44.01 \text{ g } CO_2}{1 \text{ mol } CO_2} = 66.0 \text{ g } CO_2$$

Self-Test 2

1. Carbon dioxide and water are produced when 0.250 mole of liquid ethanol [$C_2H_5OH(\ell)$] reacts with 0.150 mole of oxygen. Write and balance an equation for this combustion reaction, and identify the limiting reactant.

2. Consider the reaction of iron and oxygen to form iron(III) oxide. What is the limiting reactant if 1.300 moles of iron are placed in a reaction vessel with 64.0 grams of oxygen?

Products: Less Product Produced than Is Stoichiometrically Expected

If we consider again the reaction of butane and oxygen, that is,

$$2\,C_4H_{10}(g) + 13\,O_2(g) \rightarrow 8\,CO_2(g) + 10\,H_2O(g)$$

and react 1 mole of butane with 6.5 moles of oxygen, we would theoretically expect to produce 4 moles of carbon dioxide and 5 moles of water. However, in real-world applications, these theoretical product quantities are rarely achieved, generally for any of several reasons. One is the possibility of side reactions that consume butane or oxygen. If this occurs, not all of the initial quantities of reactants will be available for the primary reaction, reducing the amounts of products produced. A second possibility is that the primary reaction itself will not go to completion, again reducing the amounts of products produced. A third possibility is loss of product due to the experimental technique.

In these cases, we can calculate the theoretical yield of product from stoichiometry. The *theoretical yield* is the maximum amount of product that the reaction can produce. The amount of product actually produced is called the *actual yield*. The *percent yield* is given by

$$\%\text{ yield} = \frac{\text{actual yield}}{\text{theoretical yield}} \times 100\%$$

As an example, consider the gas-phase reaction of nitrogen and hydrogen to form ammonia. The balanced equation is

$$N_2(g) + 3\,H_2(g) \rightarrow 2\,NH_3(g)$$

If 42.0 g N_2 reacts with excess hydrogen to form 35.7 g NH_3, let's determine the theoretical yield and the percent yield.

Finding the theoretical yield is simply a matter of performing a standard stoichiometry calculation:

$$42.0\text{ g }N_2 \times \frac{1\text{ mol }N_2}{28.02\text{ g }N_2} \times \frac{2\text{ mol }NH_3}{1\text{ mol }N_2} \times \frac{17.03\text{ g }NH_3}{\text{mol }NH_3} = 51.1\text{ g }NH_3$$

The percent yield is found by applying the percent concept:

$$\%\text{ yield} = \frac{\text{actual yield}}{\text{theoretical yield}} \times 100\% = \frac{35.7\text{ g}}{51.1\text{ g}} \times 100\% = 69.9\%$$

Self-Test 3

1. If 1.00 g of butane reacts with excess oxygen under certain conditions, 2.81 g of carbon dioxide forms. What is the percent yield of carbon dioxide?

Thinking about Problem Solving

The flowchart on the next page illustrates one strategy you can use for stoichiometric calculations. Note that the actual process is more flexible than the flowchart indicates, and you may find that, with experience, you can combine steps.

Problem-Solving Flowchart

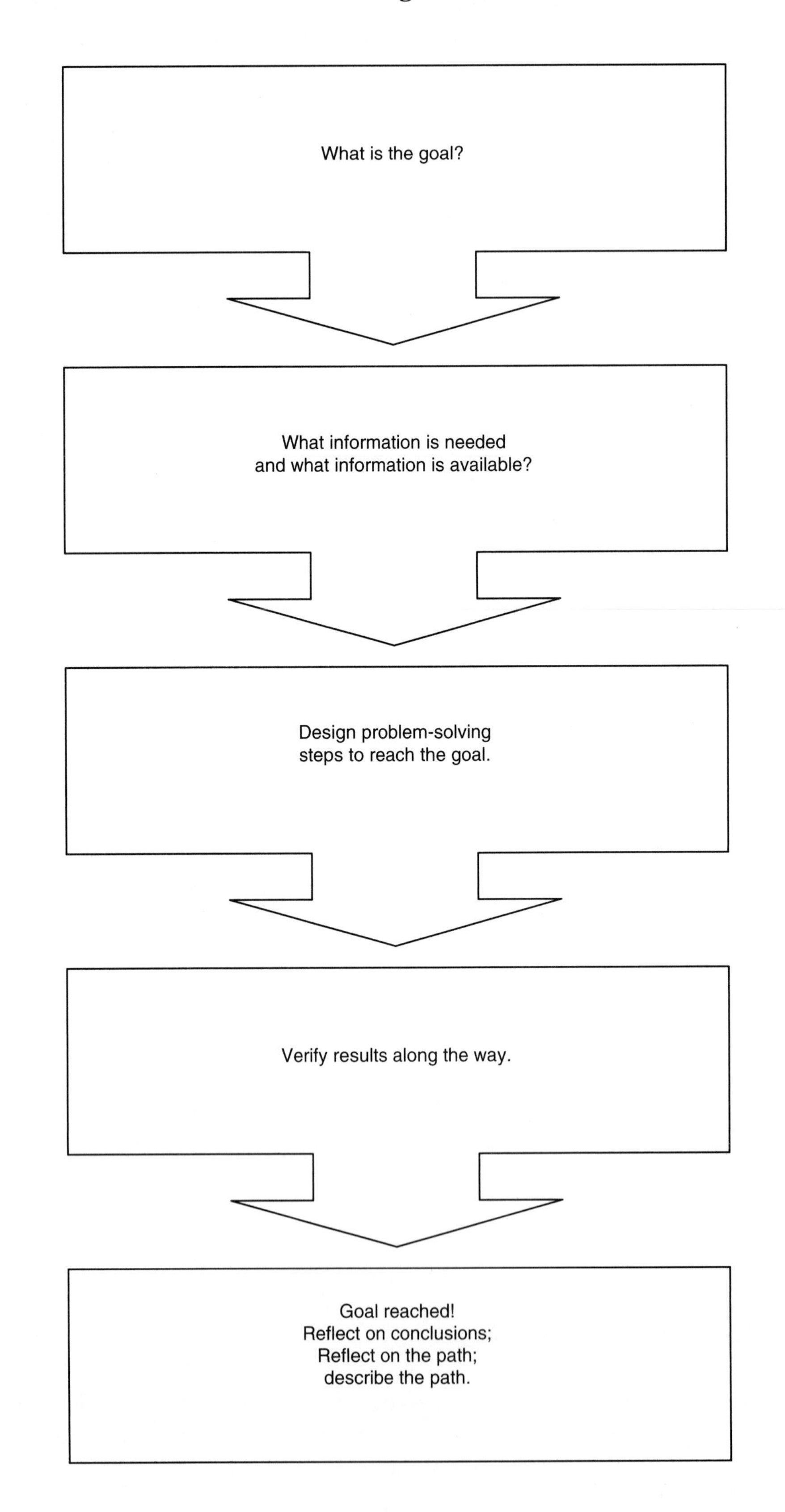

Workshop: Strategies for Stoichiometry

Work in pairs on Question 1. One member should read the problem aloud and the other should solve it, expressing his or her thinking along the way. Reverse roles for each part of the question.

1. Use coins to model a chemical reaction. You will need 15 pennies to represent oxygen atoms and 15 dimes to represent hydrogen atoms. Two dimes stacked on one another will represent a hydrogen molecule (H_2); two stacked pennies will symbolize an oxygen molecule (O_2). Two dimes and a penny in a stack will represent a water molecule (H_2O).

   a. Write and balance the equation for the reaction of hydrogen and oxygen to form water.

   b. Use your coins to show the number of water molecules that will be formed from the reaction of four hydrogen molecules and two oxygen molecules.

   c. How many water molecules will result from the reaction of six hydrogen molecules and five oxygen molecules? Verify your answer with your models. How many of which reactant molecule(s) will be left over?

   d. What is the limiting reactant when four oxygen molecules and seven hydrogen molecules react? What is the excess reactant? How many water molecules form? Use your models to verify your answers.

   e. What quantity of water will result if 4 moles of hydrogen react with 2 moles of oxygen? How does your answer to this question compare with your answer to part b? Explain.

Use the group round-robin method for Question 2.

2. Aluminum cookware is often referred to as "anodized aluminum." The anodizing process puts a layer of aluminum oxide on the aluminum, protecting it from corrosion. Consider the reaction of 12.50 grams of aluminum with 12.50 grams of oxygen to form aluminum oxide. How many grams of product are formed? Use the following framework to answer this question:

   a. What is the stated goal of the problem?

   b. What information is available and what information is needed?

   c. What is the formula of aluminum oxide? Write and balance the chemical equation for the process.

   d. Calculate the number of moles of aluminum in 12.50 g Al.

   e. Calculate the number of moles of oxygen in 12.50 g O_2.

   f. How many moles of product would be expected (i) if oxygen is the limiting reactant and (ii) if aluminum is the limiting reactant? Which is in fact the limiting reactant?

   g. Determine the number of moles of product formed.

   h. Determine the number of grams of product formed. Is the number of grams of product formed consistent with mass balance?

   i. Compare the process of solving this problem with the problem-solving flowchart on page 46. What information was needed but not provided? Which steps verify calculations or conclusions?

For Question 3, split your group into two subgroups. When both subgroups have completed the question, have each exchange and critique the other's problem-solving steps.

3. a. Solid tetraphosphorus decoxide (P_4O_{10}) reacts with liquid water to form an aqueous solution of phosphoric acid. How many grams of phosphoric acid will form when 30.00 grams of tetraphosphorus decoxide reacts with 75.00 mL of water? Assume that the density of water is 1.000 g/mL.

b. After you have completed Question 3a, compare your problem-solving method with the problem-solving process outlined on page 46. Write a description of your method that could be used as a textbook summary on how to answer this type of question.

Use the group round-robin method for Question 4.

4. Freon-12 is a gas that can be used as a refrigerant, but its use is being phased out because it causes harm to the earth's ozone layer. Freon-12 is prepared by the following unbalanced reaction:

$$\_\_\_\_ CCl_4(\ell) + \_\_\_\_ SbF_3(s) \rightarrow \_\_\_\_ CCl_2F_2(g) + \_\_\_\_ SbCl_3(s)$$

If the percent yield of freon-12 for this reaction is 72.0%, determine the number of grams of SbF_3 that must be treated with an excess of CCl_4 to obtain 25.0 grams of freon-12.

a. Balance the reaction.

b. Given that the actual yield for this reaction is 25.0 grams and the percent yield is 72.0%, what is the theoretical yield?

c. How many moles of freon-12 could be theoretically produced?

d. How many moles of SbF_3 must react to obtain the theoretical yield calculated in Part c?

e. How many grams of SbF_3 are equivalent to the number of moles calculated in Part d?

For Question 5, split your group into two subgroups. When both subgroups have completed the question, have each exchange and critique the other's problem-solving steps.

5. Hydrogen chloride is prepared commercially by the reaction of solid sodium chloride with a concentrated solution of sulfuric acid, yielding solid sodium hydrogen sulfate and hydrogen chloride gas. The percent yield of hydrogen chloride in one commercial manufacturing plant is 81.5%. How many grams of hydrogen chloride will be obtained when 25.0 kg of sodium chloride reacts with excess sulfuric acid?

Ions in Solution

> *"Our blood even contains roughly the same percentage of salt as the ocean,*
> *where the first life forms evolved. They eventually brought onto the land*
> *a self-contained store of sea water to which we are still connected, chemically and biologically."*
> *Al Gore, former vice president of the United States*

Early ideas of atoms and compounds, developed primarily through the reactions of solids and gases, did not include the concept of charge. Atoms and molecules were seen as neutral particles. However, as the study of chemistry progressed to include solutions, new models were needed because the old models could not explain electrical conductivity. Studies of the electrical conductivity of solutions and of other properties of solutions, such as freezing-point depression and osmotic pressure, showed an interesting dichotomy: On the one hand, solutions of compounds like sugar did not increase the electrical conductivity of water, yet they had lower freezing points than pure water. On the other hand, solutions of compounds such as sodium chloride greatly affected the electrical conductivity of water, and they also caused the freezing point of the solution to be reduced twice as much as was observed in sugar-water solutions of the same molar concentration.

A new model that explained these observations was based on the concept that separated, charged particles, called *ions,* could exist in solutions. If compounds like sodium chloride broke apart into charged particles when in solution, the ions could carry electrical current. Substances such as sugar must not break into ions in solution because they did not conduct electricity. These studies of the electrical characteristics of solutions led to a more complete and accurate understanding of chemistry at the particulate, or submicroscopic, level.

Solvent and Solute

When a solid dissolves in a liquid to form a solution, the solid is called the *solute* and the liquid is called the *solvent.* This is the only case that we will consider in this workshop unit. Note, however, that your instructor may require you to learn some of the other applications of these terms. For instance, a more general definition of *solute* is "the solution component that is present in the smaller amount," and a *solvent* is the component present in the greatest amount.

Solubility

Solubility is a measure of how much solute can dissolve in a given amount of solvent. A wide variety of units can be used for this purpose, although solubility is most frequently expressed in grams of solute per 100 mL of solvent at a particular temperature.

In describing how much of a given solute dissolves in water, our most common solvent, we will use the terms *soluble, slightly soluble,* and *insoluble.* Quantitative descriptions of these terms are applied quite loosely, however, and there is a great deal of variation within these categories. The rules that follow summarize the solubility characteristics of many common ionic compounds.

Simple Solubility Rules for Ionic Salts in Water

1. Most nitrate (NO_3^-) salts are soluble.
2. Most sodium, potassium, and ammonium (Na^+, K^+, NH_4^+) salts are soluble.
3. Most chloride (Cl^-) salts are soluble. Notable exceptions are $AgCl$, $PbCl_2$, and Hg_2Cl_2.
4. Most sulfate (SO_4^{2-}) salts are soluble. Combinations with sulfate ion that form insoluble compounds are $SrSO_4$, $BaSO_4$, Hg_2SO_4, Ag_2SO_4, and $PbSO_4$.
5. Most hydroxide (OH^-) salts are only slightly soluble. The important soluble hydroxides are $NaOH$, KOH, and $Ca(OH)_2$.
6. Most sulfide (S^{2-}), carbonate(CO_3^{2-}), and phosphate (PO_4^{3-}) salts are only slightly soluble.

Electrolytes and Nonelectrolytes

An *electrolyte* is a compound whose aqueous solution contains ions. When NaCl dissolves in water, the compound dissociates into Na^+ and Cl^- ions. A good test to determine whether a compound is an electrolyte is to measure the ability of its water solution to conduct an electrical current. Consider a battery, which has both a positive and a negative pole. If the poles are immersed in a solution via conductive metal electrodes, such as copper wires, the positively charged sodium ions in the solution will move toward the negative pole and the negatively charged chloride ions will move toward the positive pole. Such a solution has a high conductivity.

In contrast, if a neutral molecule such as sugar is in solution, it will not move toward either pole and the solution will be a nonconductor.

Strong and Weak Electrolytes

Electrolytes can be further classified as *strong* or *weak*. Strong electrolytes are compounds such as NaCl, which are nearly 100% dissociated in solution. This means that nearly every sodium chloride formula unit exists as sodium ions surrounded by water molecules and chloride ions surrounded by water molecules. We can represent this state of affairs by the following equation:

$$NaCl(s) \rightarrow Na^+(aq) + Cl^-(aq)$$

The Proton in Chemistry

Acids and bases form a special and very important class of electrolytes. Some acids, such as hydrochloric acid (HCl) almost completely dissociate in aqueous solution. These acids are strong electrolytes that are similar to sodium chloride, in that they exist as ions in solution. Other acids, such as acetic acid (CH_3COOH) dissociate only slightly when dissolved in water. These acids are classified as weak electrolytes. For example, at a certain concentration and temperature, only 4 of every 100 acetic acid molecules will ionize in solution. We represent the ionization of weak electrolytes in solution with double arrows as follows:

$$CH_3COOH(aq) \rightleftharpoons CH_3COO^-(aq) + H^+(aq)$$

By contrast, in the case of a strong acid such as HCl, a single arrow is used in the reaction equation:

$$HCl(aq) \rightarrow H^+(aq) + Cl^-(aq)$$

The single arrow indicates that essentially all the HCl molecules dissociate. Acids can be defined as substances that release hydrogen ions [$H^+(aq)$] in solution. The concentration of $H^+(aq)$ in solution is an important factor in a great number of chemical processes, including many that are of biological interest. Common acids that you may be familiar with include hydrochloric acid (sometimes called muriatic acid), which is used to control the acidity of swimming pools; sulfuric acid, found in automobile batteries; and phosphoric acid, which is widely used in colas for flavoring.

Now we will consider the chemical "opposite" of acids, which are compounds known as bases. A base is a compound that produces hydroxide ions [$OH^-(aq)$] in solution. As with acids, bases can be classified as either weak or strong. An example of a strong base is sodium hydroxide:

$$NaOH(s) \rightarrow Na^+(aq) + OH^-(aq)$$

Ammonia is a common weak base:

$$NH_3(aq) + H_2O(\ell) \rightleftharpoons NH_4^+(aq) + OH^-(aq)$$

When acids and bases react with each other, they form an ionic salt and water in what is called a neutralization reaction. Examples include the following:

$$NaOH(aq) + HCl(aq) \rightarrow Na^{+}(aq) + Cl^{-}(aq) + H_2O(\ell)$$

$$Ba(OH)_2(s) + H_2SO_4(aq) \rightarrow BaSO_4(s) + 2\,H_2O(\ell)$$

The salt may be soluble in water, as is sodium chloride, or it may precipitate out as a solid, as does barium sulfate.

Self-Test 1

1. For each of the following compounds, (a) write the formula and (b) classify it as either a strong or a weak electrolyte.

Compound	Formula	Strong or Weak Electrolyte
Potassium hydroxide		
Acetic acid		
Sodium chloride		
Octane		
Sucrose		
Ethanol		

Molarity: The Chemist's Favorite Concentration Unit

Chemists use a mole-based system of concentration units to measure the amount of solute in a solution. This system allows chemists to easily extend stoichiometric calculations to reactions that occur in solution. *Molarity* is defined as the number of moles of solute per liter of solution and is given the symbol M:

$$M \equiv \frac{\text{moles of solute}}{\text{volume of solution in L}}$$

When chemists prepare solutions, they usually refer to the molarity of the compound dissolved in the solution, regardless of whether it does or does not exist as ions. For example, if sufficient water is added to dissolve 1.0 mole of NaCl and bring the total volume of the solution to 1.0 L, then the solution is called a 1.0-M NaCl solution. We know that there are no NaCl particles in the solution, but rather $Na^{+}(aq)$ and $Cl^{-}(aq)$ aqueous ions.

Self-Test 2

1. In each row of the table that follows, a description of a solution is provided. (a) Calculate the molarity of the solution, and then (b) write the formulas of the major species present in the solution. The first line is completed as an example.

Solution	Molarity	Major Species in Solution
0.1000 mole of NaCl in 1.000 L of aqueous solution	0.1000 M	$Na^+(aq)$, $Cl^-(aq)$, and $H_2O(\ell)$
1.250 g of NaCl in 1.500 L of aqueous solution		
1.325 moles of acetic acid in 1.300 L of aqueous solution		
0.235 grams of NaOH in 100.0 mL of aqueous solution		

Chemical Analysis by Titration

Reactions in solution are useful for determining the amount of a particular chemical species present in a given aqueous sample. For example, swimming-pool water is often analyzed for its acid content. One way to determine the amount of acid in a solution is to titrate the solution with a base of known concentration. To analyze a solution that contains the acid HCl, for example, you can add a *known concentration* of the base NaOH in small amounts until all of the acid is neutralized. By measuring the volume of NaOH solution needed to neutralize the HCl, the number of moles of NaOH added can be determined. Since the reaction of HCl and NaOH occurs in a one-to-one ratio, at the equivalence point of the titration the number of moles of NaOH added must be equal to the number of moles of HCl in solution. The reaction equation is

$$HCl(aq) + NaOH(aq) \rightarrow Na^+(aq) + Cl^-(aq) + H_2O(\ell)$$

In every titration, we need a way to determine the point at which the reaction is complete. In the case of our sample titration of NaOH into HCl, hydrogen and hydroxide ions combine to form water molecules. This change is shown by the following equation, called the *net ionic equation* for the reaction:

$$H^+(aq) + OH^-(aq) \rightarrow H_2O(\ell)$$

The equivalence point is reached when all of the $H^+(aq)$ ions in the HCl solution have reacted. The consumption of $H^+(aq)$ can be detected by employing a chemical dye known as an *indicator*. Indicators change color when the concentration of hydrogen ions in a solution changes through the appropriate pH range for a given indicator. The color change signals the *end point* of the titration. A pH meter can also be used to measure the $H^+(aq)$ in solution and signal the equivalence point.

As we noted earlier, we can apply stoichiometric calculations to reactions that occur in solution. The macroscopic–submicroscopic conversion is made with the molarity concentration unit. Molarity allows us to convert from moles to liters and vice versa. Continuing with the titration of hydrochloric acid with a sodium hydroxide solution, let's assume that we want to know the concentration of a 25.0-mL (0.0250-L) sample of HCl. From the balanced chemical equation, we know that 1 mole of HCl reacts with 1 mole of NaOH. If the titration required 17.9 mL (0.0179 L) of 0.122-M NaOH solution, the concentration of the HCl solution is calculated as follows:

$$0.0179 \text{ L} \times \frac{0.122 \text{ mol NaOH}}{\text{L}} \times \frac{1 \text{ mol HCl}}{1 \text{ mol NaOH}} = 0.00218 \text{ mol HCl}$$

$$\frac{0.00218 \text{ mol HCl}}{0.0250 \text{ L}} = 0.0872 \frac{\text{mol HCl}}{\text{L}} = 0.0872 \text{ M HCl}$$

Self-Test 3

1. How many milliliters of 0.1000-M NaOH are required to neutralize 0.2200 L of a 0.1500-M HCl solution?

Workshop: Ions in Solution

Work in pairs to solve this problem, and then share your results with the whole group.

1. An application of the solubility concept is the identification of ions in solution by selective precipitation of their insoluble compounds. If there are several metal ions in solution, they can be identified by finding reactants that will precipitate them one at a time.

   For this question, assume that you have the following three solutions at your disposal: $NaCl$, Na_2SO_4, and $NaOH$. Your team's task is to make a flowchart showing the sequence of steps that you would use to separate the ions in each solution by precipitating each ion one at a time.

   a. Solution A contains Ag^+, Ba^{2+}, and Fe^{3+}.

   b. Solution B contains Pb^{2+}, Sr^{2+}, and Ni^{2+}.

Use the round-robin approach to this question.

2. An environmental chemist collects a 0.4546-g sample of waste material from an industrial process that releases benzoic acid ($HC_7H_5O_2$) plus additional inert compounds that are harmless to the environment. The solid sample is dissolved in 50.00 mL of water, in which it completely dissolves. The resulting solution required 10.10 mL of 0.1550 M NaOH for complete neutralization.

   a. Write the titration reaction. Benzoic acid is monoprotic, releasing one hydrogen ion per molecule.

   b. How many moles of NaOH were needed to titrate the sample?

   c. How many moles of benzoic acid were present in the dissolved sample?

   d. How many grams of benzoic acid were in the sample?

   e. What is the mass percent of benzoic acid in the sample?

Work in pairs to solve this problem, and then share your results with the whole group.

3. Each of the entries in the table that follows represents a selected point in the titration of 50.00 mL of a 0.1000-M HCl solution with 0.2000 M NaOH. Complete the table, keeping notes and showing calculations in the space below the table. In the first row, which we have completed as an example, the concentrations of H^+ and Cl^- are

$$50.00 \text{ mL} \times \frac{1 \text{ L}}{1000 \text{ mL}} \times \frac{0.1000 \text{ mol H}^+ \text{ and Cl}^-}{\text{L}} = 0.005000 \text{ mol H}^+ \text{ and Cl}^-$$

Volume OH^- Added (mL)	Moles of H^+ in Solution	Moles of Cl^- in Solution	Moles of Na^+ in Solution	Moles of OH^- in Solution	Comments
0.00	0.005000	0.005000	0	0	Calculation shown
12.50					
25.00					
30.00					

Work in pairs on this exercise. Share your sketches with the entire group.

4. Consider, once again, the titration described in Question 3. First, draw a sketch of a HCl solution and of a NaOH solution. Next, use objects such as pennies and nickels to model the reaction of hydrogen ions and hydroxide ions as the titration progresses through the four volumes listed in the table in Question 3. For example, you may wish to use pennies to represent the number of moles of H^+ in solution. You could start with 50 pennies to represent the 0.005000 mol H^+ initially in the solution. A similar scale could be used with nickels representing hydroxide ions. Show the appropriate number of hydroxide ions (nickels) added when 12.50 mL of the sodium hydroxide solution is titrated, and show also how they react with the hydrogen ions initially in the beaker. Continue the process through each step.

   When you complete your modeling exercise, draw sketches representing the particles to illustrate what occurs during this acid–base titration. Use at least one sketch for each of the four volumes listed in the table in Question 3.

Questions 5–7: Use the group round-robin approach.

5. Calculate the number of moles and grams of methanol (CH_3OH) that must be added to 2.00 mol of water to make a solution that has equal numbers of methanol and water molecules. How many total molecules does the resulting solution contain?

6. Vinegar is a 5.0% acetic acid solution [$CH_3COOH(aq)$] by mass. What is the molarity of vinegar? You may assume that the density of this dilute solution of vinegar is 1.00 g/mL.

7. How many moles of 0.100-M NaOH are required to react completely with 1.00 L of the vinegar in Question 6? How many L of NaOH is this?

Gases: Nature, Laws, and Applications

> *The important thing in science is not so much to obtain new facts as to discover new ways of thinking about them" (Sir William Bragg, 1862–1942)*

It took chemists a long time to understand the nature of gases. Maybe this was because experiments with gases required more sophisticated equipment than experiments with solids or liquids. Once scientists such as the Irish chemist Robert Boyle (1627–1691), the French chemist Jacques Charles (1746–1823), and the Italian chemist Amedeo Avogadro (1776–1856) observed and quantified the *macroscopic gas properties* of *mass, pressure, volume,* and *temperature,* scientists began to understand what a gas must be like at the *submicroscopic,* or *particulate, level.*

The Gas Laws and the Ideal Gas Equation

The pioneering studies of Boyle, Charles, and Avogadro led to a set of equations that today are expressed by the ideal gas equation. All of the other gas equations can be derived from this single equation, so it is important to understand it well.

At low pressures (less than 1 atmosphere) and high temperatures (greater than 0°C), most gases obey the *ideal gas equation*:

$$PV = nRT$$

In this equation,
P = pressure, measured in atmospheres
V = volume, measured in liters
n = amount of gas, measured in moles
T = absolute temperature, measured in kelvins
R = the *ideal gas constant*, which has a value of 0.0821 L • atm/mol • K

The ideal gas equation expresses the relationships among the pressure, volume, temperature, and amount of a gas. It provides the link between the observable macroscopic properties of pressure, volume, and temperature and a submicroscopic quantity: the number of gas particles or moles.

Example 6.1

How many grams of oxygen are in a 20.0-L tank at 27°C when the oxygen pressure is 15.0 atm?

Solution

Since mass in grams is directly proportional to moles, we can use the ideal gas law to calculate the number of moles of oxygen. The molar mass of oxygen is then used to convert to grams. We have

$$n = \frac{PV}{RT} = \frac{(15.0\text{ atm})\ (20.0\text{ L})}{(0.0821\text{ L}\cdot\text{atm/mol}\cdot\text{K})\ (27 + 273)\text{K}} = 12.2\text{ mol}$$

$$12.2\text{ mol }O_2 \times \frac{32.00\text{ g }O_2}{\text{mol }O_2} = 3.90 \times 10^2\text{ g }O_2$$

Self-Test 1

1. A balloon has a volume of 1.35 liters at 297 K and 1.05 atm.

   a. How many moles of gas are in the balloon?

   b. If the balloon is filled with helium, what is the density of the gas (g/L)?

2. An average pair of human lungs contains about 3.5 L of air after inhalation and about 3.0 L after exhalation. Assuming that air in your lungs is at 37°C and 1.0 atm, determine the number of moles of air in a typical breath.

The Origins and History of the Ideal Gas Law

When Boyle and Charles studied the properties of gases in the 17th and 18th centuries, they investigated the relationship between two of the four variables in the ideal gas equation and held the other two constant. Their work gave chemists a set of simplified gas laws upon which the ideal gas law is based.

Boyle's Law (1662)

By keeping the temperature and the number of moles of gas constant, Boyle discovered that the pressure and volume of a gas are inversely proportional. This means that, as the pressure on a confined gas increases, its volume decreases. When two variables are inversely proportional, their product is a constant. Boyle's law is expressed by the equation

$$PV = \text{constant}$$

Charles's Law (1787)

By keeping the pressure and the number of moles of gas constant, Charles discovered that the volume and temperature of a gas are directly proportional. This means that, as the temperature of a confined gas increases, its volume also increases. When two variables are directly proportional, their ratio is constant. Charles's law is expressed by the equation

$$\frac{V}{T} = \text{constant}$$

Avogadro's Law (1811)

At a given temperature and pressure, the volume and amount (number of moles, n) of a gas are directly proportional. In the form of an equation,

$$\frac{V}{n} = \text{constant}$$

Notice how the time line develops. The gas laws evolved over a period of nearly 150 years. During that time, scientists were working to understand the submicroscopic nature of matter, and Avogadro made an important contribution to that work by recognizing that a macroscopic property of gases (volume) is directly proportional to a submicroscopic property: the moles (number of particles) of gas in the sample.

To standardize the literature on experiments with gases, scientists have adopted a convention whereby gas properties are often expressed at a set of conditions known as *standard temperature and pressure* (STP), defined as a pressure of exactly 1 atmosphere and a temperature of 273.15 K. Under these conditions, the volume of 1 mole of any gas is 22.4 L.

The Kinetic–Molecular Theory of Gases

The major points of the kinetic-molecular theory are stated in the next paragraph. Textbooks often express these points differently, but they usually summarize the same major points. Find the section of your book on kinetic–molecular theory, study it, and compare it to the following summary:

The kinetic–molecular theory of gases states that a gas consists of a large number of submicroscopic particles (molecules) that are in constant random motion with a distribution of speeds. The gas particles are separated by distances that are large compared with their size and exert no forces on each other except during collisions. A gas molecule moves in a straight line until it strikes another gas particle or the wall of the container in which the gas is confined. When gas molecules collide with each other or the walls of their container, they change direction, but do not lose kinetic energy. The collisions of gas molecules with the walls of the container and with each other are *elastic* (i.e., no kinetic energy is lost in the collision).

Self-Test 2

1. What physical conditions are specified by the term STP?

2. What is the volume of 0.500 mol oxygen at STP?

3. If the oxygen gas in Question 2 is in an expandable container and the temperature is doubled at constant pressure, what volume will the gas occupy at the higher temperature?

4. If the oxygen gas in Question 2 is in an expandable container and the pressure is doubled at constant temperature, what volume will the gas occupy at the higher pressure?

5. If the oxygen gas in Question 2 is in an expandable container and 2 moles of helium are added to the container with no change in pressure and temperature, what volume will the gas mixture occupy?

6. Write a summary of the kinetic–molecular theory of gases, as presented in your textbook.

Workshop: Gases: Nature, Laws, and Applications

Questions 1 and 2: Solve the problems in pairs and then share your solution with the group.

1. A balloon is filled with helium gas at 27°C and 1.00 atm pressure. As the balloon rises, its volume increases by a factor of 1.60 and the temperature decreases to 15°C. Assuming that no helium escapes from the balloon, what is the final pressure?

2. The following statement describes one aspect of the kinetic–molecular theory of gases:

   The volume of all of the molecules of a gas is small with respect to the total volume in which the gas is contained.

   a. Compare this statement with the statements about kinetic–molecular theory in your textbook. Is there a comparable statement in your text? Explain the difference, if any.

   b. The volume of 1 mole of gaseous nitrogen at standard temperature and pressure is 22.4 L. The density of liquid nitrogen at –196°C is 0.808 g/mL. What is the volume of 1 mole of liquid nitrogen? What percentage of the volume of gaseous nitrogen is occupied by the molecules themselves?

   c. Compare the volume of 1 mole of liquid nitrogen calculated in part b with the volume of gaseous nitrogen at STP. Use the statement given at the beginning of this question to explain the difference in volume.

Questions 3 and 4: Use the group round-robin approach.

3. A gas is known to be one of the following nitrogen oxides: NO, NO_2, N_2O_4, and N_2O. It has a density of 1.96 g/L at 273 K and 1.00 atm. Determine its identity by finding its molar mass.

4. Consider the following description of an automobile air bag:
"In a frontal impact of sufficient severity, the air bag sensing system on the vehicle will detect that the vehicle is suddenly stopping as a result of a crash. The sensing system completes an electrical circuit, triggering a chemical reaction of the sodium azide sealed in the inflators. The reaction produces nitrogen gas, which inflates the air bag."
(Source: *1995 Saturn Owner's Manual*, p. 33).

The reaction that occurs is

$$2\ NaN_3(s) \rightarrow 2\ Na(s) + 3\ N_2(g)$$

How many grams of sodium azide are needed to produce 40.0 L of nitrogen to fill an air bag at a pressure of 1.30 atm and a temperature of 28.0°C?

Solve Question 5 in pairs and then compare your solution with that of the other members of your group.

5. The graphs that follow show how certain properties of the atmosphere vary with altitude (the distance above the earth's surface). Figure 1 shows how the average temperature changes with altitude. Figure 2 shows how atmospheric pressure changes with altitude. Figure 3 shows how the molar mass of air changes with altitude. Use the graphs and your knowledge of the ideal gas law to calculate the density of air, in g/L, at altitudes of 5 km and 10 km.

Figure 1:
Altitude vs. Average Temperature

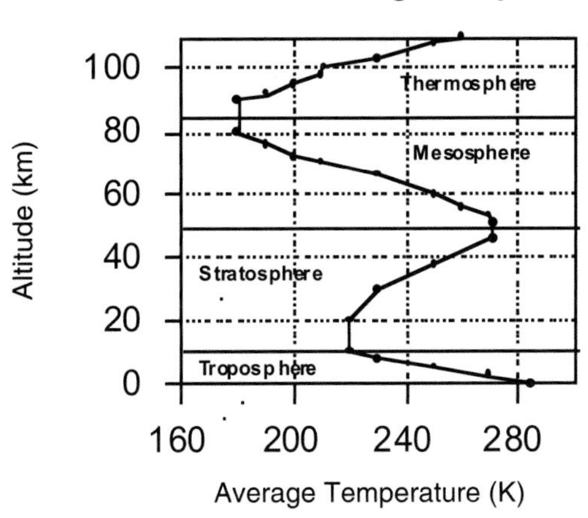

Figure 2:
Altitude vs. Pressure

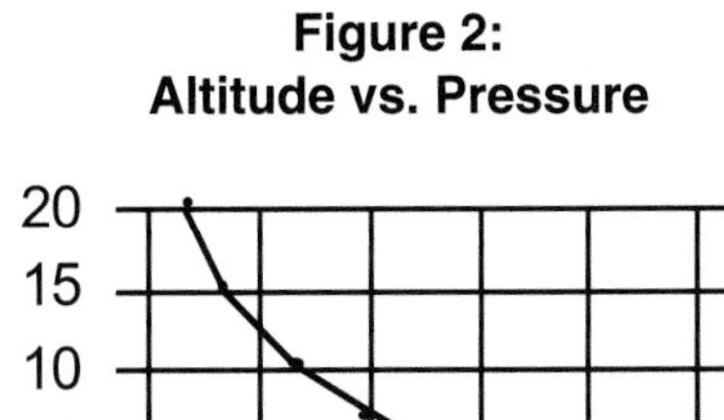

Figure 3:
Altitude vs. Average Molar Mass of Air

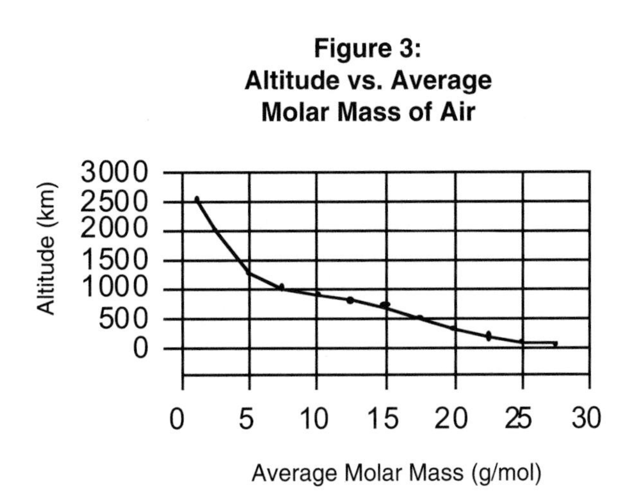

Figures adapted from Stanley R. Radel and Marjorie H. Navidi, *Chemistry,* 2nd ed. (Minneapolis–St. Paul: West, 1994), p. 197.

Discuss Question 6 as a group. Explain the pros and cons of each possible combination.

6. Consider the following possible combinations of hydrogen and oxygen to form water:

i. $H + O \rightarrow HO$

ii. $2\,H + O \rightarrow H_2O$

iii. $H_2 + O \rightarrow H_2O$

iv. $2\,H + O_2 \rightarrow 2\,OH$

v. $H_2 + O_2 \rightarrow 2\,OH$

vi. $2\,H_2 + O_2 \rightarrow 2\,H_2O$

a. Dalton believed that hydrogen and oxygen were monoatomic gases. Which equation(s) would Dalton have considered consistent with his hypothesis? Explain.

b. Which equation(s) would Avogadro have considered consistent with his hypothesis that equal numbers of molecules occupy equal volumes and that hydrogen and oxygen were diatomic gases? (Recall that 2 volumes of hydrogen react with 1 volume of oxygen to form 2 volumes of water.)

Work in pairs to simulate the chain reaction in Question 7. Summarize the work of the group on the board when the entire group reaches a consensus.

7. Our present understanding of atmospheric chemistry leads us to believe that the earth's protective ozone layer is suffering destruction because of the release of human-made chlorofluorocarbons (CFCs) into the atmosphere. As a result, the United States and a number of other nations have banned or agreed to phase out the use of CFCs. They are still being used, however, in a number of developing countries. $CFCl_3$, one of the most common CFCs, reacts with light energy, symbolized here as $h\nu$, to dissociate into reactive free radicals—chemical species with unpaired electrons—according to the following equations (the dots in the formulas indicate the species with unpaired electrons):

$$CFCl_3(g) + h\nu \rightarrow CFCl_2\cdot(g) + Cl\cdot(g)$$

$$Cl\cdot(g) + O_3(g) \rightarrow ClO\cdot(g) + O_2(g)$$

$$ClO\cdot(g) + O(g) \rightarrow O_2(g) + Cl\cdot(g)$$

This reaction sequence depicts a chain reaction—that is, a reaction that is self-sustaining. The chain reaction occurs because the chlorine radical, $Cl\cdot(g)$, is regenerated by reacting with oxygen atoms in the atmosphere.

Use your molecular model kit to simulate a chain reaction. Let a red atom represent oxygen and a green atom represent chlorine. (Alternatively, use pennies to represent oxygen and nickels to represent chlorine.) To simulate the reaction, start with five ozone molecules (three oxygen atoms per molecule), five oxygen molecules (two oxygen atoms per molecule), and one chlorine atom. Show how a single chlorine atom can react with all five ozone molecules. (*Hint*: Oxygen atoms are formed by the light-induced dissociation of the O_2 molecule.)

Use the group round-robin approach for Question 8.

8. Amidst all the laws and equations we have learned in this unit, the importance of the kinetic–molecular theory of gases to our understanding of matter may be lost.

   a. Although Boyle discovered the pressure–volume relationship experimentally, it can be derived from kinetic molecular theory. Using your summary of that theory from Question 6 in Self-Test 2, describe qualitatively how the theory can be used to account for Boyle's law.

   b. How does kinetic–molecular theory explain Charles's law?

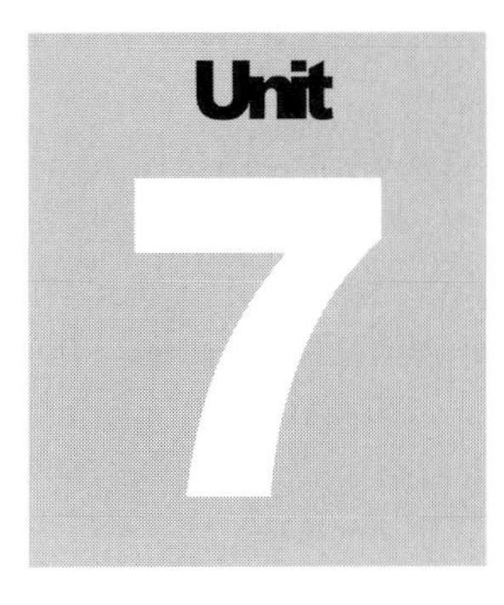

Thermochemistry

> *"You can't win."*
> *Casey Stengel*

Thermochemistry is the application of the laws of thermodynamics to chemical processes. In this unit, we will be concerned with the first law of thermodynamics and its applications. You are probably already familiar with the first law, because it is the same as the law of conservation of energy: Energy can be neither created nor destroyed; it can only be converted among different forms.

System and Surroundings

When we perform an experiment designed to test a hypothesis related to a thermochemical process, or when we think about thermochemistry, we divide the universe into two parts, respectively called the system and the surroundings. The *system* is the part of the universe under consideration. The *surroundings* are the remainder of the universe. In a general chemistry laboratory, thermochemical experiments are often conducted in water in a styrofoam cup. The heat released from the physical or chemical change is absorbed by the water and the cup (the system). In many thermochemical experiments, it is desirable to minimize the amount of heat energy that can transfer between the system and the surroundings (everything outside of the cup). As coffee drinkers know, styrofoam cups do a good job of keeping the transfer of heat energy relatively low compared with other containers, such as paper cups. In more sophisticated research settings, more expensive instruments (calorimeters) are used that provide better insulation and reduce the transfer of heat energy further.

State Functions

A *state function* is a property of a system that depends only on its present condition or state. Some properties of gases, such as pressure, temperature, and volume, are state functions. The value of a state function is constant for a specified set of properties that describe the state. For example, if a sample of an ideal gas has a specific volume, amount, and temperature, the value of the pressure must be constant for the given set of properties.

The difference between the values for two different sets of properties that determine the value of a state function is also constant. Thus, the difference in pressure between an ideal gas at one specific set of volume, amount, and temperature conditions and the same gas at a different set of volume, amount, and temperature conditions is a fixed value. This illustrates an important idea about state functions: They do not depend on the history of how they changed. If you know the initial values of V, n, and T for an ideal gas system and the final values of V, n, and T for the same system, then you also know the change in pressure, no matter what the values of V, n, and T were during the change in the gas from its initial to its final state.

Energy

The *internal energy* of a system, E, is a state function. The change in internal energy, ΔE, is the sum of the heat flow into or out of the system, q, and the work done on or by the system, w:

$$\Delta E = q + w$$

The conventions used to establish the arithmetic signs for heat flow and work are important to understand. You must first carefully determine the identity of the system; once that is done, the following conventions apply:

If heat is *gained* by the system, q is positive.
If heat is *lost* by the system, q is negative.
If work is done *on* the system, w is positive.
If work is done *by* the system, w is negative.

Enthalpy

Another important state function is the *enthalpy* of a system, defined as

$$H \equiv E + PV$$

where E is internal energy, P is pressure, and V is volume. Note that all of the functions that define enthalpy are themselves state functions. Enthalpy, therefore, must also be a state function.

The change in enthalpy is given by the enthalpy of the final state minus the enthalpy of the initial state, which, for a chemical reaction, is products minus reactants:

$$\Delta H = H_{products} - H_{reactants}$$

If $H_{products}$ is greater than $H_{reactants}$, the change in enthalpy is positive and the process is endothermic. If $H_{products}$ is less than $H_{reactants}$, the change in enthalpy is negative and the process is exothermic.

The change in internal energy of a system can also be expressed in terms of the definition of enthalpy:

$$\Delta H = \Delta E + \Delta(PV)$$

Most chemical changes occur at constant pressure, so we have

$$\Delta H = \Delta E + P\Delta V$$

Substituting the sum of heat flow and work for the change in internal energy ($\Delta E = q + w$) yields

$$\Delta H = q + w + P\Delta V$$

At constant pressure, work is defined as $w = -P\Delta V$, and substituting this term, we have

$$\Delta H = q + (-P\Delta V) + P\Delta V$$

which leaves

$$\Delta H = q$$

at constant pressure. To indicate that this relationship is valid only at constant pressure, a subscript p is included in the expression:

$$\Delta H = q_p$$

The quantity q_p is the heat flow at constant pressure. This relationship tells us that, for the common condition of constant pressure, change in enthalpy and heat flow are equal. Thus, q, which could be difficult to measure because it is not a state function, can be found by finding ΔH, which is a state function; therefore, only the initial and final states need to be known to find ΔH and q for a constant-pressure process.

Self-Test 1

1. Write the definition of each of the following: (a) state function, (b) enthalpy, (c) system, (d) surroundings, (e) heat, and (f) temperature.

Thermochemical Equations

A *thermochemical equation* is a balanced chemical equation that includes the change in enthalpy for the reaction. The physical state—gas, liquid, solid, or aqueous—must be included for each chemical species in the reaction, because enthalpy varies for the same substance in different physical states. As an example, consider the combustion of liquid ethanol:

$$C_2H_5OH(\ell) + 3\,O_2(g) \rightarrow 2\,CO_2(g) + 3\,H_2O(\ell) \quad \Delta H = -1367 \text{ kJ}$$

The change in enthalpy change is –1367 kJ per 1 mole of ethanol or per 3 moles of oxygen that react. Similarly, 1367 kJ of energy is released per 2 moles of carbon dioxide or per 3 moles of water that form as products.

If 2 moles of ethanol react with excess oxygen, the enthalpy released is

$$2 \text{ mol } C_2H_5OH \times \frac{1367 \text{ kJ}}{1 \text{ mol } C_2H_5OH} = 2734 \text{ kJ}$$

Self-Test 2

1. Calculate the change in enthalpy when 20.2 g of oxygen reacts according to the equation
$$H_2(g) + \tfrac{1}{2}\,O_2(g) \rightarrow H_2O(g) \quad \Delta H^\circ = -242 \text{ kJ}$$

2. Calculate ΔH°_f for liquid water, using the thermochemical equation given in Question 1 and the thermochemical equation
$$H_2O(\ell) \rightarrow H_2O(g) \quad \Delta H^\circ = +44 \text{ kJ}$$

Standard State

The change in enthalpy for a chemical reaction is dependent on both temperature and pressure. Chemists have therefore established a standard temperature and pressure at which thermochemical quantities are to be reported whenever it is practical to do so. These *standard state conditions* are 298.15 K (25°C) and 1 bar (10^5 Pa) pressure. In addition, the most stable form of the substance is included as a standard state condition. When a change in enthalpy is at standard state, it is indicated by the symbol $\Delta H°$.

The thermodynamic standard state for pressure recently was redefined as 1 bar. Previously, it was 1 atm (1.013×10^5 Pa). The difference between these two pressures is very slight, 1.3%, and it has little consequence in most thermochemical calculations. If your instructor prefers to use 1 atm as standard pressure, we recommend that you follow his or her advice.

The Law of Heat Summation

Given that change in enthalpy is a state function, we can determine its value for a specified reaction by the summation of a series of reactions that are algebraically equal to the reaction of interest. This approach is particularly useful when it is difficult to measure the change in enthalpy for a reaction. The *law of heat summation* can be stated mathematically as

$$\Delta H_{\text{overall reaction}} = \Delta H_{\text{reaction}_1} + \Delta H_{\text{reaction}_2} + \ldots$$

where the overall reaction is obtained by adding reactions 1, 2, ...

As an example, consider the reaction of graphite and oxygen to form carbon monoxide, whose change in enthalpy is difficult to measure experimentally:

$$C(s, \text{graphite}) + \tfrac{1}{2}\, O_2(g) \rightarrow CO(g)$$

Two other reactions for which changes in enthalpy are known are

$$C(s, \text{graphite}) + O_2(g) \rightarrow CO_2(g) \qquad \Delta H° = -393.5 \text{ kJ}$$

and

$$CO(g) + \tfrac{1}{2}\, O_2(g) \rightarrow CO_2(g) \qquad \Delta H° = -283.0 \text{ kJ}$$

We can reverse the carbon monoxide–plus–oxygen reaction and add the reactions:

$$\begin{array}{llr} C(s, \text{graphite}) + O_2(g) & \rightarrow CO_2(g) & \Delta H° = -393.5 \text{ kJ} \\ CO_2(g) & \rightarrow CO(g) + \tfrac{1}{2}\, O_2(g) & \Delta H° = +283.0 \text{ kJ} \\ \hline C(s, \text{graphite}) + \tfrac{1}{2}\, O_2(g) & \rightarrow CO(g) & \Delta H° = -110.5 \text{ kJ} \end{array}$$

Note that when a reaction equation is reversed, the sign of the enthalpy change is opposite of that for the forward reaction.

Standard Enthalpies of Formation

The *standard molar enthalpy of formation* of a substance, ΔH_f°, is the change in enthalpy required to form 1 mole of the substance from its elements in their standard states. For any element in its standard state, ΔH_f° is zero. For example, graphite is the most stable form of carbon at 298 K and 1 atm. From this definition, we can measure ΔH_f° for a compound by producing it from its elements. For example, if we wish to determine ΔH_f° for carbon dioxide, we would measure the change in enthalpy for the reaction

$$C(s, \text{graphite}) + O_2(g) \rightarrow CO_2(g)$$

This reaction results in a change in enthalpy of –393.5 kJ. Thus, we state that ΔH_f° for $CO_2(g)$ is –393.5 kJ/mol.

The standard change in enthalpy for a reaction can be determined from the standard enthalpies of formation of the species in the reaction. For any reaction,

$$\Delta H^\circ = \sum \Delta H_f^\circ{}_{\text{products}} - \sum \Delta H_f^\circ{}_{\text{reactants}}$$

Self-Test 3

1. Using ΔH°_f values from your textbook, calculate ΔH° for the reaction of $CH_4(g)$ with oxygen gas to produce gaseous carbon dioxide and liquid water.

Thermochemical Measurements

Specific heat is the amount of heat needed to raise the temperature of 1 gram of a pure substance by 1°C. It can be determined by measuring the change in temperature as a known quantity of a substance gains or loses a measured quantity of heat. The formula is

$$c \equiv \frac{q}{m\Delta T}$$

where c is the specific heat. A calorimeter is used to study heat flows and specific heats.

Self-Test 4

1. How much heat is absorbed by 52.3 grams of water if the sample was originally at 22.6°C and was heated to 35.8°C? The specific heat of liquid water is 4.18 J/g °C.

Workshop: Thermochemistry

1. Split your group into two subgroups. One subgroup will write the definition of the first three terms listed in the table that follows, and the other will write the definition of the other three terms. Once the definitions are finished, discuss them until the entire group is satisfied that they are correct and complete. Next, write examples to illustrate each term. The subgroup that wrote the definitions of the first three terms should give examples for the second set of terms, and vice versa. When the examples are complete, the whole group should discuss both the definitions and the examples until everyone in the group fully understands all of the terms.

Term	Definition	Example
Thermochemical equation		
Standard-state enthalpy of reaction		
Specific heat		
Stoichiometric chemical equation		
Standard-state enthalpy of formation		
Heat capacity		

Questions 2 and 3: Use the paired problem solving approach. Compare your results with those from other pairs.

2. Consider the following reaction and its associated enthalpies of formation:

$$CS_2(\ell) + 3\,O_2(g) \rightarrow CO_2(g) + 2\,SO_2(g)$$

Species (state)	$\Delta H°_f$ (kJ/mol)
$CS_2(\ell)$	+ 87.9
$CO_2(g)$	– 393.5
$SO_2(g)$	– 296.8

a. Determine the standard-state enthalpy of reaction.

b. Construct a one-dimensional plot using a vertical axis to represent increasing enthalpy. On this axis, draw a short horizontal line to indicate the enthalpies of each of the species in the reaction, the sum of the enthalpies of the reactants, and the sum of the enthalpies of the products. How does $\Sigma\,\Delta H°_{f\ reactants}$ compare with $\Sigma\,\Delta H°_{f\ products}$? Is the reaction endothermic or exothermic? How would your plot change if the reactants and products of the reaction were reversed?

3. Calculate ΔH for the reaction

$$N_2O(g) + NO_2(g) \rightarrow 3\ NO(g)$$

from the following enthalpies of reaction:

$$NO(g) + \tfrac{1}{2}\ O_2(g) \rightarrow NO_2(g) \qquad \Delta H = -56.6\ kJ$$
$$N_2O(g) \rightarrow N_2(g) + \tfrac{1}{2}\ O_2(g) \qquad \Delta H = -81.6\ kJ$$
$$\tfrac{1}{2}\ N_2(g) + \tfrac{1}{2}\ O_2(g) \rightarrow NO(g) \qquad \Delta H = +90.4\ kJ$$

Use the group round-robin approach for Question 4.

4. Consider the reaction

$$Ag(s) + \frac{1}{2}Cl_2(g) \rightarrow AgCl(s) \quad \Delta H° = -127 \text{ kJ}$$

a. Is the reaction endothermic or exothermic? How do you know?

b. Calculate the amount of heat transferred when 50.5 g of solid silver reacts at constant pressure.

c. Calculate the mass of solid silver chloride formed when 25.8 kJ of heat is released.

d. What quantity of heat is needed to decompose 102.5 g of solid silver chloride into its elements?

Discuss Question 5 as a group.

5. Without looking up the actual values, predict the relative specific heats of solid aluminum (aluminum foil), liquid water, and steam. Justify your answer with reasoning based on your everyday experiences. After you have a ranking and a justification, look up the actual specific heats of these substances in your textbook. If there are any discrepancies, try to figure out why.

Solve Question 6 in pairs and then compare your approach with that of the others in your group.

6. A 6.58-g sample of solid sodium hydroxide is dissolved in 127.3 g of water in a calorimeter. What will be the change in temperature of the solution? The heat of solution for solid sodium hydroxide is 44.5 kJ/mol.

Use the paired problem-solving approach to Question 7, switching roles after completion of the first part. Compare the approaches of each pair as a group.

7. A student conducts a calorimetry experiment to determine the specific heat of an unknown metal. She nests two polystyrene cups in one another. She then adds 50.0 mL of room-temperature water at 20.5°C to the cup. She warms water in a beaker, and then she adds 50.0 mL of water at 60.9°C to the water already in the cup. She records the temperature of the water as it changes. Through graphical extrapolation, she determines that the final temperature of the water is 38.1°C. She then empties and dries her cup. She again adds 50.0 mL of water at 20.5°C. Her 58.98-g sample of the unknown metal was sitting in a boiling-water bath during the first part of the experiment. The temperature of the water was 99.2°C. She quickly transfers the metal to the cup and records the water temperature again. This time, the temperature is 27.3°C. What is the specific heat of the metal?

   Break the described experiment into two parts. Explain why each part was performed, and explain what quantities, if any, can be calculated from the given information. Your final calculation will be the specific heat of the metal. When you complete the question, go back and summarize the procedure used to solve this type of problem.

Unit

8

Energy and the Hydrogen Atom

> *"The 'silly question' is the first intimation of some totally new development."*
> *Alfred North Whitehead*

The idea that energy comes in discrete "chunks," called quanta, was developed in the early 20th century by Niels Bohr and Albert Einstein. Quantum theory, the result of understanding that energy comes in discrete packets, has very important consequences for chemistry. To understand the quantum mechanical model of the atom, it is necessary to understand the concepts *force* and *energy*—particularly the nature and behavior of radiant energy. This workshop will concentrate on these concepts and on the idea of quantized energy in the hydrogen atom.

Electromagnetic Radiation

Historically, a study of the emission spectrum of hydrogen led to a deeper understanding of the structure of the hydrogen atom. Today, light energy continues to hold a very important role in chemistry. Many important modern methods of chemical analysis, including those of great practical use in medicine and technology, involve the interaction of light energy with matter.

Visible light, infrared and ultraviolet radiation, microwaves, radio waves, and X-rays are all forms of radiant energy that are known collectively as *electromagnetic radiation*. The *electromagnetic spectrum* is the range of wavelengths and frequencies that scientists recognize as electromagnetic radiation. The term "electromagnetic" is used because this radiation consists of oscillating electric and magnetic fields. It is the interaction of the oscillating electric field with the electrons in atoms and molecules that is of interest to chemists.

Properties of Waves When electromagnetic radiation travels through a vacuum, it has properties similar to those of an ocean wave traveling through water. A wave is characterized by three quantities: wavelength (λ), frequency (ν), and amplitude.

Wavelength is the distance between any two adjacent points of a wave. *Frequency* is the number of wavelengths that pass through a fixed point in a unit of time, usually 1 second. *Amplitude* is the wave height, measured from its middle point to its peak or trough. The following diagram illustrates wavelength and amplitude:

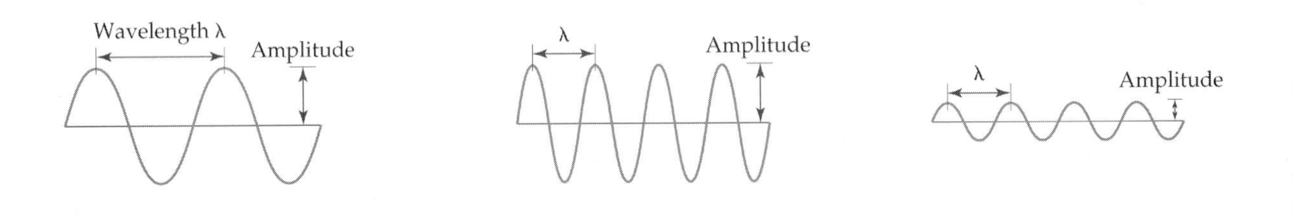

Table 8.1 lists some types of electromagnetic radiation and their wavelengths. Notice the wide range of wavelengths found in the electromagnetic spectrum. Radio waves can be as long as 1000 meters, while gamma rays can be as short as 0.0000000000001 meter.

Energy of Electromagnetic Radiation As you know, ultraviolet radiation from the sun can be harmful to your skin. People who spend more than a few minutes outside are advised to use a sunscreening lotion that blocks the sun's harmful rays. Because of its high frequency and short wavelength, ultraviolet radiation is a highly energetic form of electromagnetic radiation. Overexposure to ultraviolet rays can cause skin cancer.

The energy of an electromagnetic wave is related to its frequency, wavelength, and speed. All forms of electromagnetic radiation travel at a constant speed in a vacuum: 3.00×10^8 m/s. The relationships described next will enable you to determine the energy of any type of electromagnetic radiation.

Type of Radiation	Frequency (Hz)	Wavelength
Radio	10^6 to 10^8	1 km to 30 cm
Microwave	10^9 to 10^{12}	30 to 2 mm
Infrared	10^{12} to 10^{14}	2 mm to 700 nm
Visible	10^{14}	700 nm to 400 nm
Ultraviolet	10^{14} to 10^{17}	400 nm to 4 nm
X-rays	10^{17} to 10^{19}	4 nm to 30 pm
Gamma rays	larger than 10^{19}	30 pm to 0.1 pm

Table 8.1. The Electromagnetic Spectrum

The product of wavelength (λ) and frequency (ν) is equal to the speed of the wave, c:

$$\lambda\nu = c$$

Since c is constant, wavelength and frequency are inversely proportional. This means that short waves have high frequencies and long waves have low frequencies.

The energy (E) carried by a wave is equal to the product of Planck's constant, which has a value of 6.626×10^{-34} Js, and the frequency (ν) of the wave:

$$E = h\nu$$

This equation states that the energy carried by an electromagnetic wave is directly proportional to the wave's frequency.

If the equation $\lambda\nu = c$ is solved for frequency, we have

$$\nu = \frac{c}{\lambda}$$

If we then substitute this expression for frequency into the equation $E = h\nu$, we obtain

$$E = \frac{hc}{\lambda}$$

Now we can directly calculate the energy of a wave from its wavelength. Since h and c are both constant, the preceding equation tells us that energy is inversely proportional to wavelength.

Self-Test 1

1. Which has a longer wavelength, electromagnetic radiation with a frequency of 10^4 Hz or electromagnetic radiation with a frequency of 10^7 Hz?

Wave–Particle Duality of Light and Matter

In 1905, Albert Einstein proposed that light has both wave and particle properties. He reasoned that light is composed of a stream of particles known as *photons*, each with energy given by $E = h\nu$. Einstein received the Nobel Prize in physics in 1921 for his ideas about the particle nature of light. Soon thereafter, in 1924, a French graduate student, Louis de Broglie, suggested that, since light can behave as a particle, then perhaps matter can have wavelike properties. Modern scientists have validated these ideas experimentally. Light behaves like waves in some cases and like particles in other cases; and moving objects, such as electrons, also behave like both waves and particles. We refer to this phenomenon as *wave–particle duality*.

In 1926, an Austrian scientist, Erwin Schrödinger, devised a quantum mechanical model of the atom that used complex wave equations to describe electrons. Solutions of Schrödinger's wave equation yield results known as wave functions, which describe the volume in space where an electron with a particular energy is likely to be found. This volume in space is called an *orbital.* Schrödinger's wave equation also explains the line spectra of multielectron atoms and the arrangement of electrons in these atoms. Quantum mechanics is explored in Unit 9.

Self-Test 2

1. The frequency of radio station WFAN is 660 kHz.

   a. What is the wavelength (m) of the radio waves emitted by WFAN?

   b. What is the energy (kJ) of a photon emitted by WFAN?

   c. What is the energy (kJ) of 1 mole of these photons?

2. What is the frequency (Hz) of a radar wave with λ = 10.3 cm?

Force and Energy

Chemists base their understanding of chemical reactions and atomic structure on concepts from physics, such as force, energy, acceleration, and heat. This section briefly reviews the concepts of force and energy.

Weight and Gravity Weight is the force acting on an object because of the gravitational attraction of the earth. For objects on the surface of the earth, the gravitational force can be expressed by the equation

$$F = mg$$

where m is the mass of the object and $g = 9.8\ m/s^2$, the acceleration due to gravity.

The gravitational force of the earth can be related to the general definition of force as something that causes an object to accelerate—that is, to increase its velocity. The general equation for force is

$$F = ma$$

where a is acceleration.

Force Units When mass is measured in kilograms and acceleration is measured in m/s^2, the unit of force is the newton (N). One newton is the force that increases the velocity of a 1-kilogram object by 1 meter per second in 1 second.

Kinetic Energy Kinetic energy is energy of motion. The motion can be from point to point, in which case it is called translation. The motion can also be movement around a fixed axis, called rotation or vibration. For simple translation, kinetic energy is given by

$$KE = \frac{1}{2} mv^2$$

where m is the mass of the object and v is its velocity.

Potential Energy Potential energy, or energy of position, is a result of forces acting at a distance. Examples of potential energy include the force of gravitation between masses such as the sun and the earth and the force of attraction between particles having opposite charges.

Self-Test 3

1. At what point do the objects that follow achieve their greatest velocity? Explain how potential and kinetic energy vary during each object's flight.
   a. A rock falling to the ground
   b. A baseball thrown from a pitcher's hand

2. A diver jumps from the high board into a pool. Sketch an illustration of the diver's motion. Label each of the following on your sketch:
   a. The position of maximum potential energy
   b. The position of maximum kinetic energy

Potential Energy of Charged-Particle Interaction With only one proton and one electron, hydrogen is the simplest atom. The energy of the electron in a hydrogen atom is the sum of its potential energy and the kinetic energy coming from the proton–electron interaction. In 1913, Niels Bohr, a Danish scientist, proposed a planetary model of the hydrogen atom in which electron motion is similar to the motion of planets in our solar system as they move around the sun. Let's begin an analysis of this system of charged-particle interaction with a consideration of the potential energy.

The potential energy of an interaction between two charged particles is given by

$$PE = \frac{kq_1q_2}{r}$$

where k is a constant, q represents the charge on the interacting particles, and r is the distance between the particles. For a hydrogen atom, the values for k and q are as follows:

$$k = 8.98755 \times 10^9\ \mathrm{N \cdot m^2/C^2}$$

$$q_{electron} = -\ 1.602 \times 10^{-19}\ \mathrm{C}$$

$$q_{proton} = +\ 1.602 \times 10^{-19}\ \mathrm{C}$$

Since potential energy is energy of position, we must choose a reference point to measure it. Scientists have established the convention of using a distance of infinity as the reference point for potential energy. Look at the formula and decide what happens when r becomes very large—as r approaches infinity. You can see that potential energy must become very small and eventually go to zero as r goes to infinity.

Returning to our consideration of the hydrogen atom, we find that the potential energy of the electron will become more negative—its potential energy decreases—as it moves closer to the proton. The electron's potential energy approaches zero—its potential energy increases—as it gets farther away from the proton.

Energy Units The SI unit for energy is the joule, which has units of kg m^2/s^2. A 1-kilogram object accelerated to a velocity of 1 m/s in 1 s has a kinetic energy of 1 joule.

Total Energy Any system of interacting objects has an energy that, at least theoretically, can be calculated. The total energy of the system is equal to the sum of the potential and kinetic energies:

$$\text{TE} = \text{PE} + \text{KE}$$

Since the total energy of the system is constant, the change in total energy is given by

$$\text{Change in TE} = \text{Change in PE} + \text{Change in KE} = 0$$

If we use the Greek uppercase delta (Δ) to represent the change in any given quantity, we can express the change in total energy as

$$\Delta\text{TE} = \Delta\text{PE} + \Delta\text{KE}$$

Atomic Spectra and Energy Levels

When elements are heated, individual atoms emit visible light. The light that is emitted is limited to a few specific colors and is called a *line spectrum.* Each element has its own unique line spectrum. For example, the visible portion of hydrogen's line spectrum consists of four colors: indigo (λ = 410.1 nm), blue (λ = 434.0 nm), green (λ = 486.1 nm), and red (λ = 656.3 nm). Line spectra are used to identify elements and to confirm the discovery of new elements.

Bohr developed the planetary model of the hydrogen atom while trying to understand the emission spectrum of hydrogen. According to Bohr's model of the hydrogen atom, the electron moves around the proton in a circular orbit. When it absorbs energy, it moves to a higher orbit, one that is farther from the nucleus. Conversely, when it drops back to a lower orbit, it releases energy in the form of radiant energy, some of which is visible light.

The Bohr model of the atom does not take into account wave–particle duality. It did successfully explain hydrogen's line spectrum, however. Bohr also determined that the energies of the allowed orbits could be determined by the equation

$$E = -R_H \left(\frac{1}{n^2}\right)$$

where E is the energy of the electron in orbit n, with values of n ranging over the numbers 1, 2, 3, ..., and R_H is a constant known as the Rydberg constant, which depends on the mass of the electron, the charge of the nucleus, and Planck's constant. The value of R_H for the hydrogen atom is 2.18×10^{-18} J.

In addition to explaining the line spectrum of hydrogen, the Bohr model predicted that the radius of the electron orbit with $n = 1$ would be 5.3×10^{-11} m, which was exactly the radius of the hydrogen atom determined by other means. Bohr's model also predicted that the energy needed to remove the electron from the hydrogen atom—the ionization energy—would be 2.18×10^{-18} J, which was in excellent agreement with the experimentally determined value. In addition, Bohr's model correctly predicted the line spectra of hydrogenlike ions that have only one electron, such as He^+ and Li^{2+}.

Still, while the Bohr model correctly explained the properties of the hydrogen atom, it could not account for atoms or ions with more than one electron. Also, the model did not explain *why* energy should be quantized. Two subsequent discoveries solved these problems and led to the quantum mechanical model of the atom, which today serves as the foundation for chemistry and the basis of the periodic table. To understand the quantum mechanical model of the atom and the concept of quantized energy, it is necessary to examine the nature of light energy and the electromagnetic spectrum.

Workshop: Energy and the Hydrogen Atom

Questions 1–3: Solve in pairs, and then compare your solutions with those of the entire group.

1. Mercury vapor lamps, which are used for street and highway lighting, produce light at a lower overall cost than the incandescent lamps that they have replaced. One of the lines in the spectrum of mercury is in the blue region and has a wavelength of 435.8 nm.

   a. What is the frequency of this line?

   b. What is the energy of the photon emitted?

2. The idea of a zero potential energy is very important in chemistry, applying to interactions between charged particles such as the proton and the electron in a way similar to that of gravitational potential energy. Consider again the equation for the potential energy between two charged particles:

$$PE = \frac{kq_1q_2}{r}$$

You can see that zero potential energy is approached as the distance between the objects goes to infinity.

a. Draw a vertical number line that includes all the integers from –5 to +5. Place –5 at the bottom of the line and +5 at the top.

b. Assume that the numbers represent energy levels. What is the sign on the change in potential energy (ΔPE) as you move from 0 to –5? What is ΔPE when you move from –4 to –1? Which has more energy, the zero state or a negative-numbered state?

c. If the values of k, q_1, and q_2 are something other than zero in the foregoing potential energy equation, what is the value of r when PE = 0? Explain.

d. Apply what you have learned thus far in this question to the Bohr model of the hydrogen atom. In particular, what is the sign on the potential energy of an electron in a hydrogen atom? What happens to the potential energy as an electron moves closer to the nucleus? farther from the nucleus?

3. a. Using the equation $E = -R_H \left(\frac{1}{n^2}\right)$, calculate the energies that correspond to the first through sixth principal energy levels of the electron in the Bohr hydrogen atom.

b. Calculate the energy difference between the first and fifth principal energy levels in a hydrogen atom.

c. What are the frequency and wavelength of the wave associated with the photon emitted when an electron in a Bohr hydrogen atom drops from $n = 5$ to $n = 1$?

Use the round-robin approach for Question 4.

4. Arrange the following electron transitions in the hydrogen atom in order of increasing energy change, without actually calculating all the energy changes:

a. $n = 1$ to $n = 99$

b. $n = 3$ to $n = 80$

c. $n = 2$ to $n = 3$

d. $n = 1$ to $n = 3$

e. $n = 5$ to $n = 12$

Use paired problem solving for Question 5. The entire group should discuss the results when the pairs are finished.

5. Energies of the electron in single-electron species are given by the formula

$$E = -R_H (Z^2) (\frac{1}{n^2})$$

where Z is the nuclear charge.

a. Calculate the energy difference between the first and fifth principal energy levels of the He^+ ion. How does your result compare with the energy difference for the same transition in the hydrogen atom (calculated in Question 2b)?

b. What are the frequency and wavelength of the wave associated with the photon emitted when an He^+ electron drops from $n = 5$ to $n = 1$? In what part of the electromagnetic spectrum will the line corresponding to this transition be found?

c. Compare the difference in energies between principal energy levels in the hydrogen atom, the He^+ ion, and the Li^{2+} ion. Which has the electron that makes the largest quantum jump between equivalent energy levels?

The group round-robin approach is effective for Question 6.

6. De Broglie derived the following equation for the wavelength of a particle:

$$\lambda = \frac{h}{mv}$$

Here, h is Planck's constant, m is the mass of the particle, and v is its velocity. To honor his work, the wavelength of a particle is referred to as its *de Broglie wavelength.*

a. Determine the de Broglie wavelength, in pm, of an electron in the first principal energy level of the hydrogen atom. The mass of an electron is 9.11×10^{-31} kg, and the velocity of an electron in the first principal energy level is 2.2×10^{6} m/s.

b. The electron of a hydrogen atom is usually no farther than 100 pm from the proton. We can, therefore, say that the radius of an isolated hydrogen atom is about 100 pm. How does the de Broglie wavelength of the electron compare with this radius?

c. The wavelength of visible light ranges from about 4×10^{-7} m to about 7×10^{-7} m. Microscopes using visible light can resolve details to about 10^{-6} m. Why can an electron microscope (one that uses electron waves rather than light waves) resolve details in the nanometer-to-picometer range?

d. What is the de Broglie wavelength of a 120-g baseball traveling at 95 mph (42 m/s)? Compare the baseball's wavelength and radius. What is the physical meaning of the de Broglie wavelength of a baseball?

Brainstorm as a group to come up with a plan for solving Question 7, and then use the round-robin approach to complete the solution.

7. The first ionization energy (IE) of an atom can be measured by a technique known as photoelectron spectroscopy, in which light of wavelength λ is directed at an atom, causing an electron to be ejected from the atom. The kinetic energy of the ejected electron, E_k, is measured by determining the velocity of the electron and utilizing the relationship for the kinetic energy of a particle, $E_k = \frac{1}{2}mv^2$. Finally, the IE is determined by an application of the law of conservation of energy. The energy of the light directed at the atom must be equal to the sum of the energy needed to remove the electron from the atom, which is its ionization energy, and the kinetic energy of the ejected electron.

   A photoelectron spectroscopy experiment is conducted in which light with $\lambda = 58.4$ nm is directed at a sample of rubidium, resulting in the ejection of electrons with a velocity of 2.45×10^6 m/s. Use these experimental data to determine the IE of rubidium in kJ/mol. Compare your result with the value for the IE of rubidium from your textbook.

Building Atoms with Quantum Leaps

"Oh boy . . ."
Sam Beckett (from "Quantum Leap"*)*

Physicists Put Atom in Two Places at Once

This section head was the headline in the science section of the *New York Times* on May 28, 1996. "Impossible!" you say. "How could they do that?" you wonder. This event *is* impossible at the macroscopic level at which classical mechanics governs the world. But it is entirely possible according to quantum mechanics, which governs the submicroscopic world of atoms, protons, and electrons. The physicists who caused a single beryllium atom to exist in two states at the same time, separated by a distance of 83 nanometers, made use of a quantum mechanical trait called spin.

Quantum Numbers

The spin attribute is one of the four quantum mechanical characteristics needed to describe an electron completely. Each of these characteristics is described by what is known as a quantum number. If we consider quantum numbers to make up the "address" of each electron within an atom, each address has four parts, and no two electrons have the exact same address.

The Bohr model of the hydrogen atom gave us our first clue that electrons were governed by nonclassical mechanics, and the model worked well in explaining the properties of the electron in the hydrogen atom. However, it failed for all other atoms. In 1926, Erwin Schrödinger devised a new model of the atom now known as the quantum mechanical model.

The Quantum Mechanical Model of the Atom Schrödinger's atomic model is framed mathematically in terms of what is known as a *wave equation*. Solutions of wave equations are called *wave functions*. The solutions of a wave equation define the volume in space where an electron with a particular energy is likely to be found. This volume is called an *orbital*. Each orbital is characterized by three quantum numbers.

The Pauli Exclusion Principle This principle states that no two electrons in an atom can have the same four quantum numbers. If two electrons occupy the same orbital, they must have different spins.

Following are the four quantum numbers:

1. The principal quantum number n.

   The allowed values of the principal quantum number are $n = 1, 2, 3, ..., 7$. Electrons with the same value of n are said to have the same principal energy level.

2. The angular momentum quantum number ℓ.

   Angular momentum quantum numbers depend on principal quantum numbers. For $n = 1$, $\ell = 0$. For $n = 2$, $\ell = 0$ or 1. For $n = 3$, $\ell = 0$, 1, or 2. For $n = 4$ (and higher), $\ell = 0$, 1, 2, or 3. This pattern can be summarized as $\ell = 0, 1, ..., n - 1$. Notice that all principal energy levels are divided into one or more sublevels.

   Angular momentum quantum numbers are often referred to by using letter designations that correspond to their numerical values. $\ell = 0$ is also called the *s* sublevel, $\ell = 1$ is *p*, $\ell = 2$ is *d*, and $\ell = 3$ is *f*. Electron energies are described by the principal energy level and the sublevel. Thus, an electron with $n = 3$ and $\ell = 1$ is referred to as a 3*p* electron.

3. The magnetic quantum number m_ℓ.

   Magnetic quantum numbers depend on angular momentum quantum numbers. The pattern is $m_\ell = -\ell, ..., 0, ..., +\ell$. Thus for $\ell = 0$, the only allowed value of m_ℓ is 0. When $\ell = 1$, m_ℓ can be -1, 0, or $+1$. For $\ell = 2$, $m_\ell = -2, -1, 0, +1$, and $+2$. When this pattern is followed for $\ell = 3$, there are seven possible m_ℓ values. (Can you write them?)

4. The electron spin quantum number m_s.

   The values of m_s are $+\frac{1}{2}$ and $-\frac{1}{2}$. Electrons can be thought of as spinning on an axis, where one m_s value corresponds to a clockwise rotation and the other value corresponds to a counterclockwise rotation. (Electron spin lacks an exact analog at the macroscopic level. Electrons don't really spin—at least in the way we think of ordinary macroscopic objects as spinning.)

Self-Test 1

1. Complete the table that follows by entering all possible values for the quantum numbers. The first line is completed as an example.

n	ℓ	m_ℓ	m_s
1	0	0	$+\frac{1}{2}, -\frac{1}{2}$
2			
3			
4			

2. Consider an atom of beryllium, which has two valence electrons. The quantum numbers for one of the valence electrons is $n = 2$, $\ell = 0$, $m_\ell = 0$, and $m_s = +\frac{1}{2}$. What are the quantum numbers for the other valence electron?

3. The ground-state electron configuration for the single electron in the hydrogen atom is $1s^1$. What are the values of the m_ℓ and m_s quantum numbers for this electron? Explain.

4. The quantum numbers for one of the two electrons of helium are $n = 1$, $\ell = 0$, $m_\ell = 0$, and $m_s = -\frac{1}{2}$. What are the quantum numbers for the other valence electron? Write the electron configuration for helium.

The Periodic Table

The periodic table serves as a guide to both the order of increasing electron energies and the order in which electrons fill orbitals. Electrons occupy the lowest-energy orbitals available, and as the number of electrons in an atom increases, the outermost electrons occupy higher and higher energy levels. The periodic table shown next illustrates the correspondence of electron energy levels and the position of an atom in the table.

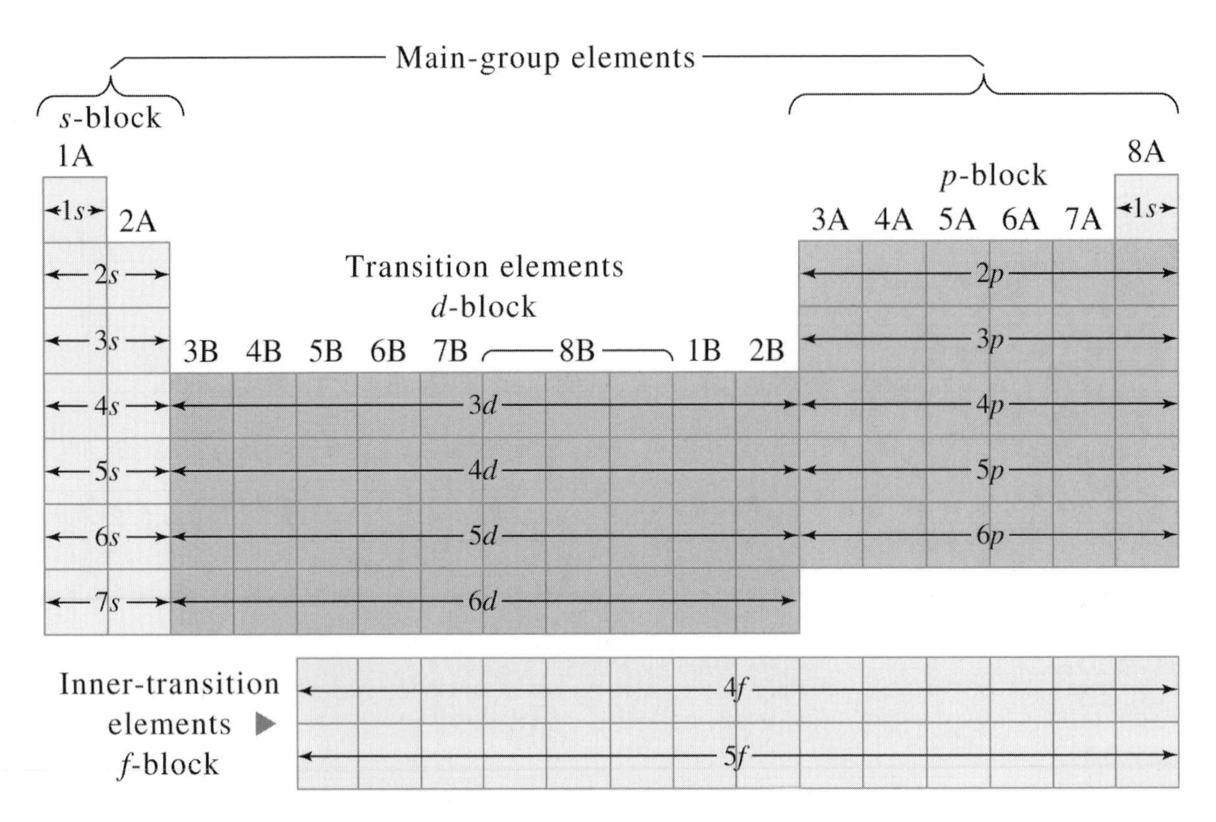

John W. Hill, Ralph H. Petrucci, Terry W. McCreary, and Scott S. Perry, *Chemistry,* 4th Edition, © 2005. Reprinted by permission of Pearson Education, Inc., Upper Saddle River, NJ.

Note:

s orbitals are being filled in Groups 1A–2A, *p* orbitals are being filled in groups 3A–8A, *d* orbitals fill in the B Groups, and *f* orbitals fill in the lanthanide and actinide series.

For *s* and *p* orbitals, the period number corresponds to the principal energy level. For *d* orbitals, the fourth period corresponds to $n = 3$, the fifth period to $n = 4$, and so on. For *f* orbitals, the *n* quantum numbers are 4 and 5. Lanthanides correspond to $n = 4$, and the actinides have $n = 5$.

Self-Test 2

1. What is the maximum number of electrons an atom can have when $n = 1$ and when $n = 2$? Explain.

Workshop: Building Atoms with Quantum Leaps

1. Count off the numbers from 1 to 36, rotating among the members of your group. These numbers correspond to the atomic numbers of the first 36 elements.

   a. Write the full ground-state electron configuration for each of your elements.

   b. Explain the relationship between the electron configuration and the corresponding principal and angular momentum quantum numbers for each of your elements.

2. Begin by assigning one of each of the following elements to each group member: carbon, nitrogen, fluorine, neon, sodium, aluminum, phosphorus, chlorine, iron. (There should be more elements in this list than group members; you may omit an element or two as necessary.) Each person should complete the following five steps for his or her element:

Step 1: Write the electron configuration of the element. This can be copied from Question 1.

Step 2: Prepare a table similar to the one that follows, giving the values of each of the four quantum numbers for each electron in the atom.

Step 3: Fill in all of the items in the table.

Step 4: Create an energy-level diagram for your element. Use a line to represent an orbital, and use up and down arrows to represent electrons in each orbital.

Step 5: Fill in the energy-level diagram.

We will complete the five steps for boron as an example.

Step 1: Electron configuration

B: $1s^2 2s^2 2p^1$

Steps 2 and 3: Chart of quantum numbers

n	ℓ	m_ℓ	m_s
1	0	0	$+\frac{1}{2}$
1	0	0	$-\frac{1}{2}$
2	0	0	$+\frac{1}{2}$
2	0	0	$-\frac{1}{2}$
2	1	–1	$+\frac{1}{2}$

Steps 4 and 5: Energy-level diagram

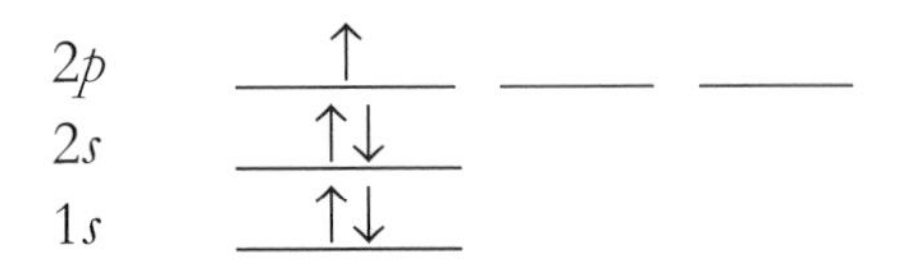

3. Answer each of the following questions in a round-robin:

a. How many 3*p* electrons does a ground-state chlorine atom have?

b. Give the complete set of quantum numbers for all of the chlorine 3*p* electrons.

c. How many *d* electrons does a ground-state iron atom have?

d. Considering the ground state, how many unpaired electrons are in a chlorine atom, an iron atom, and a nitrogen atom?

The group round-robin method works well for this question.

4. Many substances exhibit no magnetic properties, except in the presence of a magnetic field. Substances with all paired electrons are weakly repelled by a magnetic field. These substances are called *diamagnetic.* Other substances are attracted to a magnetic field and are called *paramagnetic.* This property is the result of unpaired electrons.

   Which of the following elements or ions do you expect to be paramagnetic? Explain how you reach your conclusions.

   a. K

   b. Ba^{2+}

   c. Fe^{2+}

   d. Fe^{3+}

   e. F

   f. F^{-}

   g. Ni^{2+}

5. As a group, define each of the following terms: atomic radius, ionization energy, and electron affinity. Consider the following questions in constructing your definitions:

   a. How can the radius of an atom be experimentally determined?

   b. What is meant by the first, second, and third ionization energies?

   c. What is the sign of electron affinity if it is defined as the quantity of energy released? What if it is defined as the change in energy? Which definition are you using in your class?

   d. What trends do you see in atomic radius, first ionization energy, and electron affinity as you move down a group or across a period in the periodic table?

6. Most general chemistry textbooks have a plot of first ionization energy versus atomic number. Find this figure in your textbook. What are the periodic trends in first ionization energy? Specifically, address the following questions in round-robin fashion:

 a. What is the overall trend within a group? Why?

 b. What is the overall trend within a period? Why?

 c. How do main-group elements compare with transition elements?

 d. Why are there irregularities in the trends?

7. Divide the group into teams, with each team choosing one pair of elements. For each of the pairs that follow, determine the atom with (i) the largest radius, (ii) the greatest ionization energy, and (iii) the smallest electron affinity. Report your results to the group and explain your reasoning.

a. F or K

b. O or Si

c. Li or Cs

d. F or As

Discuss Question 8 as a group.

8. What is the definition of "metallic character"? Is there a periodic trend for metallic character across a period or down a group? What is a metal? What is a nonmetal? What are the macroscopic characteristics that distinguish metals from nonmetals? What are the submicroscopic characteristics that lead to these two classes? What is a metalloid?

Covalent Bonding

> *"Never put trust in anything but your own intellect . . . always think for yourself."*
> *Linus Pauling*

Linus Pauling, perhaps the greatest chemist of the 20th century, published a landmark paper in 1931 entitled "The Nature of the Chemical Bond." His paper provided a comprehensive understanding of chemical bonding. In this workshop, your team will apply concepts that Pauling introduced to help you understand the nature of the chemical bond.

Essential Concepts

We begin with a brief review of some of the essential concepts that provide the foundation for understanding covalent bonding. The purpose of this review is to provide you with a checklist of topics that you should know before meeting with your workshop group. If any of these concepts are unfamiliar, study the appropriate section of your textbook before your workshop group meets.

Valence Electrons Electrons that can participate in bonding. These are seen to be the outermost *s* and *p* electrons when the eight-electron, or octet, rule is used.

Self-Test 1

1. Determine the number of valence electrons for each of the following elements: N, S, Cl, P, O, F, C, Si.

Lewis Dot Structures Structures drawn to represent the distribution of valence electrons in atoms and compounds.

Octet Rule The rule which asserts that elements tend to gain, lose, or share electrons in order to achieve an outer shell of eight electrons. In many compounds formed by the *p*-block elements, the total valence electron count for every atom is equal to eight. Each atom counts all electrons that it shares with other atoms, as well as its own lone-pair electrons.

Incomplete Octet The formation of covalent compounds without the acquisition of an octet of electrons. The elements beryllium, boron, and aluminum are the elements that form an incomplete octet. Lewis structures of compounds of these elements are drawn with less than an octet.

Electronegativity A quantitative measure of the relative ability of the nucleus of an element to attract electrons in a bonding pair toward itself.

Resonance The application of two or more Lewis structures to a single species. The actual bonding system is a composite of the individual structures. A species with resonance structures is said to be resonance stabilized. The bonding electrons have some freedom to move to positions of lower potential energy in resonance-stabilized species; thus, the overall species is more stable. Use caution in looking for resonance structures; specifically, electrons can change positions in resonance structures, but nuclei cannot.

Formal Charge The number of valence electrons in an isolated atom, minus all of the unshared electrons and half of the bonding electrons. Lewis structures are drawn to minimize the number of atoms with a formal charge and to place negative formal charges on the most electronegative elements and positive formal charges on the most electropositive elements.

Free Radical A species with one or more unpaired electrons. Free radicals cannot satisfy the octet rule; thus, they are highly reactive, because they readily engage in chemical reactions, producing more stable compounds with complete octets. Nitrogen compounds commonly exist as free radicals because nitrogen has an odd number of valence electrons, namely, five. Examples include NO and NO_2.

Self-Test 2

1. For each of the following compounds, draw the Lewis diagram and determine the formal charge on each atom: (a) HCl, (b) CO_3^{2-}, (c) NH_3.

Pauling Electronegativity Values for Selected Elements

Pauling developed a theoretical method for using experimentally determined data to derive a quantitative scale of electronegativity values. Following is a table of these values:

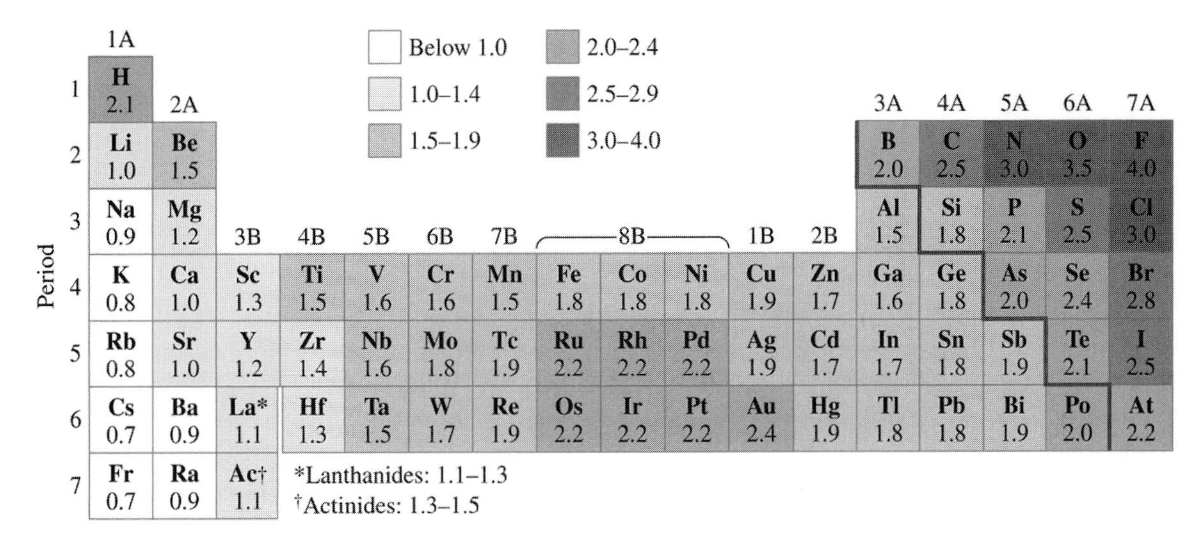

Period	1A	2A	3B	4B	5B	6B	7B	8B	8B	8B	1B	2B	3A	4A	5A	6A	7A
1	H 2.1																
2	Li 1.0	Be 1.5											B 2.0	C 2.5	N 3.0	O 3.5	F 4.0
3	Na 0.9	Mg 1.2											Al 1.5	Si 1.8	P 2.1	S 2.5	Cl 3.0
4	K 0.8	Ca 1.0	Sc 1.3	Ti 1.5	V 1.6	Cr 1.6	Mn 1.5	Fe 1.8	Co 1.8	Ni 1.8	Cu 1.9	Zn 1.7	Ga 1.6	Ge 1.8	As 2.0	Se 2.4	Br 2.8
5	Rb 0.8	Sr 1.0	Y 1.2	Zr 1.4	Nb 1.6	Mo 1.8	Tc 1.9	Ru 2.2	Rh 2.2	Pd 2.2	Ag 1.9	Cd 1.7	In 1.7	Sn 1.8	Sb 1.9	Te 2.1	I 2.5
6	Cs 0.7	Ba 0.9	La* 1.1	Hf 1.3	Ta 1.5	W 1.7	Re 1.9	Os 2.2	Ir 2.2	Pt 2.2	Au 2.4	Hg 1.9	Tl 1.8	Pb 1.8	Bi 1.9	Po 2.0	At 2.2
7	Fr 0.7	Ra 0.9	Ac† 1.1														

*Lanthanides: 1.1–1.3
†Actinides: 1.3–1.5

In general, the greater the difference in electronegativity between two bonded atoms, the greater is the ionic character of the bond. A bond between cesium and fluorine, for example, with an electronegativity difference of $4.0 - 0.7 = 3.3$ is essentially completely ionic. A bond between two identical atoms with an electronegativity difference of zero is completely covalent. Bonds in between these extremes are partly ionic and partly covalent.

Most bonds between two different nonmetal atoms have small electronegativity differences and are mostly covalent, although the electron cloud that bonds the atoms is asymmetrically distributed, favoring the more electronegative element. These bonds are referred to as *polar covalent bonds.* A nitrogen–oxygen bond, for instance, has an electronegativity difference of 0.5, and the bonding electrons will favor the oxygen end of the bond.

We can assign rough guidelines for distinguishing between "pure" covalent, polar covalent, and ionic bonding types. For "pure" covalent bonding, the electronegativity difference must be zero; for polar covalent bonding, the electronegativity difference must be greater than zero but less than 1.7; for ionic bonding, the electronegativity difference must be greater than or equal to 1.7.

Workshop: Covalent Bonding

1. Your group leader will randomly assign one or more of the steps listed next to each group member. As a group, you will then cooperatively **draw Lewis diagrams for each of the species that follow.** Each group member will answer the question(s) or follow the instructions in his or her step or steps in turn, explaining the reasoning involved to the group. After each diagram is completed, one member will summarize the entire process. Before beginning the next diagram, rotate steps by passing the assignments to the group member to the right. Each group member will have a turn at solving each of the steps.*

STEP 1: How many valence electrons are available for bonding?

STEP 2: Determine the atom-to-atom connectivity. For every atom-to-atom connection, make a single bond (a shared electron pair).

STEP 3: Distribute the remaining electrons to lone-pair positions on atoms that need an octet.

STEP 4: Identify atoms that need an octet, but do not yet have one.

STEP 5: Rearrange electrons from lone pairs of adjacent atoms, forming multiple bonds, so that the atoms identified in Step 4 have an octet. Count the electrons to make sure you have the number you started with!

STEP 6: Calculate the formal charge on each atom.

STEP 7: Are the bonds nonpolar covalent, polar covalent, or ionic? Can you draw resonance structures? If so, how many?

Use the round-robin method described in the instructions for this problem to analyze the following compounds:

a. C_2H_6

b. C_2H_4

c. C_2H_2

*Your instructor may prefer an alternative method for constructing Lewis diagrams. If so, list the steps given by your instructor on the board and follow those steps in this question.

d. NH_3

e. HBr

f. H_2O_2

g. SO_4^{2-}

h. N_2O (use an N–N–O arrangement)

i. NO_3^-

j. CO_3^{2-}

k. PCl_5

l. $BeCl_2$

m. C_6H_6 (What if the two ends of the chain connect?)

2. The three different types of chemical bond can be represented with a triangular plane as follows:*

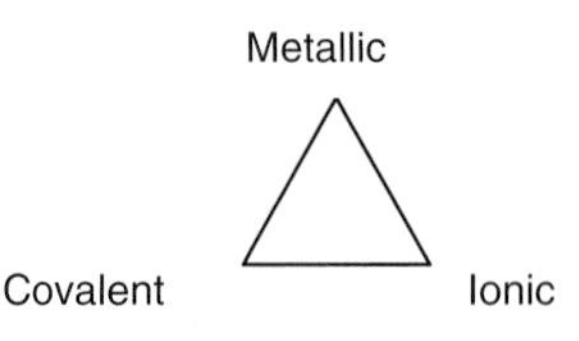

a. Define each of the three types of bond.

b. Given the name or formula of an element or a compound, how do you decide which type of bond is most likely to occur between pairs of atoms? To answer this question, consider the following: First, complete the calculations in the table that follows. The first column lists a number of compounds for which the type of bond is to be determined. The second column is for the electronegativity difference: the absolute value of the difference in electronegativity between the atoms being considered, or, symbolically, $|EN_1 - EN_2|$. The third column is for the average electronegativity of the two atoms: $(EN_1 + EN_2) \div 2$. Complete the calculations for the second and third columns now. Develop a strategy to divide the labor among the group members before proceeding.

Compound	$\mathbf{\vert EN_1 - EN_2 \vert}$	$\mathbf{\frac{EN_1 + EN_2}{2}}$	**Type of Bond**
HF			
HCl			
HBr			
HI			
CsF			
NaF			
CaO			
Na			
Fe			
Pt			
Cu			
F_2			
ClF			
BN			
Cs			

*Sproul, G. (1993). Electronegativity and Bond Type: I. Triparate Separation. *Journal of Chemical Education, 70*(7), 531–534.

c. Now construct a plot of electronegativity difference (on the y-axis) versus average electronegativity (on the x-axis). Write the formula for the element or compound next to each data point.

d. Is there a pattern in the graph? Did certain types of element or compound clump together? Which bonding type is represented by each group? Explain. Complete the fourth column of the table in Part b.

3. A *concept map* is a diagram that shows connections among concepts. Since chemical bonding is of such overall importance to chemistry, it is useful to reflect on everything you know so far about bonding. You can do this by listing all of the smaller concepts that make up the larger concept of bonding and relating them in a concept map. Here is a sample concept map for the ideal gas law:*

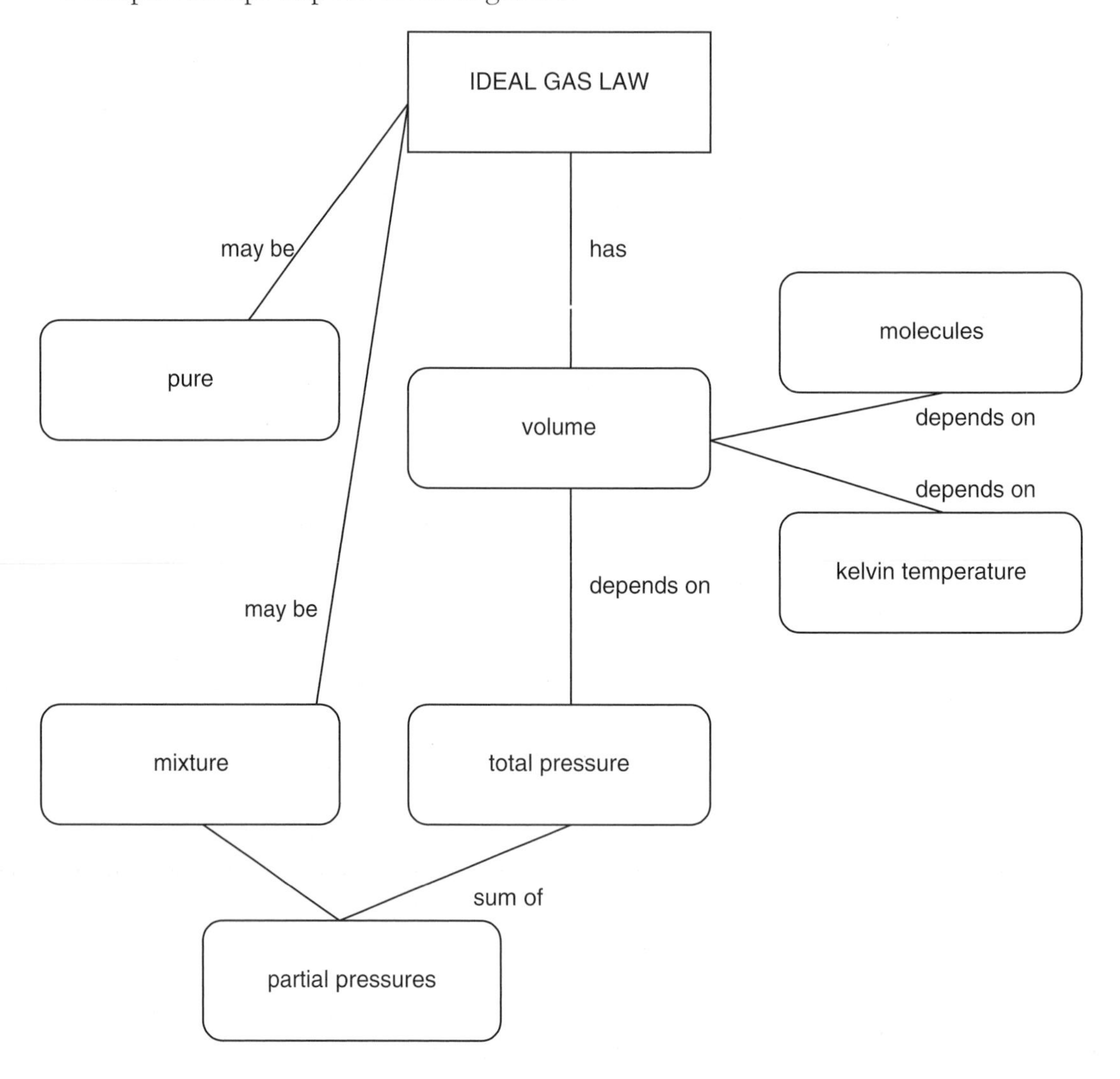

a. As a group, generate a list of 8 to 10 concepts that are critical to understanding chemical bonding. Identify one or two key concepts that you think will help link all the others.

b. Split your group into two subgroups. Each subgroup should create a concept map of chemical bonding.

c. Compare the two concept maps. Find links between them.

*This sample concept map is based on Figure 12.3 (p. 159) from Herron, J. D. (1996). *The Chemistry Classroom: Formulas for Successful Teaching.* Washington, DC: American Chemical Society.

The Structure of Molecules

> *"The carbon atom has four valence electrons. The spectrum shows that one of the electrons is different from the other three, and yet the four bonds of the carbon atom seem to be identical with one another . . . .*
> *I had the idea that the electrons might occupy four equivalent tetrahedral orbitals . . . .*
> *I worked at my desk all night, so full of excitement that I could hardly write."*
> *Linus Pauling*

Two-dimensional, or "connectivity," representations of molecules and ions help us understand and discuss the chemical bond. However, molecules are three dimensional, so we need a three-dimensional model to help us visualize the actual structure of a molecule. Three-dimensional representations also inform us about the interactions among different molecules and interactions among different parts of the same molecule.

In the workshop presented in this unit, we explore one of the models that chemists use to predict molecular structure: the valence shell electron pair repulsion (VSEPR) model. The VSEPR model is a useful starting point for learning about the three-dimensional structure of molecules, but it does not have a firm theoretical basis. In this course, you may learn about other, more rigorous models, such as the orbital hybrid model mentioned by Linus Pauling in the unit-opening quote. If so, carefully study the hybridization and/or molecular orbitals sections of your textbook in preparation for the workshop.

The Valence Shell Electron Pair Repulsion Model

The VSEPR model considers the interaction among the electrons within a molecule or ion as the determining factor of structure. The VSEPR model considers each "group" of electrons in the Lewis diagram as an arm of electron density projecting from the central atom. Each arm repels all the other arms, and the individual arms try to get as far away from each other as possible. In this context, we will consider a "group" of electrons to be one of the following:

a. a single bond (two shared electrons)
b. a double bond (four shared electrons)
c. a triple bond (six shared electrons)
d. an unshared pair (a lone pair)

The electron-pair geometry, or arrangement of electron groups around a central atom, is based on the number of groups around that atom. Each geometry has a name, which chemists use to describe the shape:

Number of Electron Groups around the Central Atom	Electron-Pair Geometry
2	Linear
3	Trigonal Planar
4	Tetrahedral
5	Trigonal bipyramidal
6	Octahedral

The molecular geometry is determined by the number of atoms bonded to the electron groups around the central atom. Your instructor may hold you responsible for knowing geometries other than those included on our list or may have you learn only a subset of those on our list. Also, sometimes terminology varies slightly. We recommend that you check with your instructor to see if there are any variations for which you will be responsible. Make any changes directly on the table:

Number of Electron Groups around the Central Atom	Number of Groups of Electrons Bonded to an Atom	Molecular Geometry
2	2	Linear
3	3	Trigonal Planar
3	2	Angular
4	4	Tetrahedral
4	3	Trigonal pyramidal
4	2	Bent
5	5	Trigonal bipyramidal
5	4	Seesaw
5	3	T shaped
6	6	Octahedral
6	5	Square pyramidal
6	4	Square planar

Each of the molecular geometries has a characteristic or ideal bond angle. Three of these are as follows:

Molecular Geometry	Ideal Bond Angle	Structure	Example
Linear	180°	 X—A—X	BeF_2
Trigonal Planar	120°	 X—A(X)(X)	BF_3
Tetrahedral	109.5°	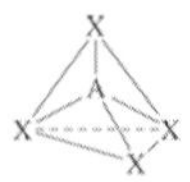 A with four X	CH_4

Self-Test 1

1. For each of the following compounds, (i) draw a Lewis diagram, (ii) count the number of electron groups around the central atom and the number of bonded electron groups, (iii) draw a three-dimensional representation of the molecule, (iv) give the values of the ideal bond angles, and (v) give the name of the electron-pair and molecular geometries:

   a. CH_4

   b. NO_3^-

   c. PCl_5

   d. SO_4^{2-}

A Refinement of the VSEPR Model

Lone electron pairs, which are not confined between nuclei, use more space than do bonding pairs, which are restricted to stay in the narrow space between their two nuclei. The "fat" lone pairs distort the ideal electron-pair geometries predicted by VSEPR theory. Consider the ammonia molecule as an example. It has four electron pairs; thus, VSEPR predicts a tetrahedral geometry with 109.5° bond angles. However, one of the four is a lone pair, taking up more space, so the three bonding electron pairs are squeezed together slightly. Thus, the experimentally measured H–N–H bond angle in ammonia is 107.3°.

Dipole Moment

A polar molecule is also known as a *dipole*. The distribution of charge is asymmetric in a dipole, resulting in positive and negative poles. In other words, there is a little more negative charge on one end of the molecule than on the other. The amount of "unevenness" in charge distribution varies among different molecules. To illustrate this concept, consider CO_2 and H_2O, which are both three-atom molecules.

Carbon dioxide has two electron groups, both bonded, so its molecular geometry is linear. Oxygen is more electronegative than carbon, so if we considered the C=O bond alone, the bonding electrons would be asymmetrically distributed toward the oxygen atom. But the bond is not alone: There are two C=O bonds in CO_2, opposite one another:

The pull toward the oxygen atom on the left exactly balances the pull toward the oxygen atom on the right. The polar bonds cancel one another. Thus, the molecule itself is nonpolar.

Water has four electron groups, with two bonded, so it has a bent molecular geometry. Oxygen is more electronegative than hydrogen, so an isolated O–H bond has its bonding electrons distributed toward the oxygen atom. Now let's consider the overall molecule:

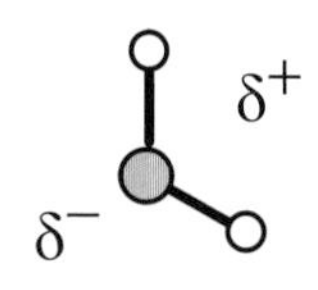

The pull toward the oxygen atom is not balanced by a pull in the opposite direction. In fact, both pairs of bonding electrons will be distributed toward the oxygen atom, creating a significant buildup of electrons on that end of the molecule. We indicate this buildup with a δ^-, which is used to symbolize a partial negative charge. The opposite end of the water molecule will be electron deficient and therefore possess a partial positive charge, indicated by the symbol δ^+.

A polar molecule such as water is said to have a dipole moment. The dipole moment of molecules can be quantitatively measured, and it is usually expressed in a unit called a debye, with symbol D. The unit is named after Peter Debye, who made many important contributions to our understanding of molecular polarity. For the purpose of this discussion, we will limit our use of quantitative dipole moments to express the relative polarity of molecules. The more polar a molecule, the greater is its dipole moment. Nonpolar molecules have a zero dipole moment.

Water, with a dipole moment of 1.85 D, has one of the largest dipole moments of any molecule. This property of water molecules leads to significant effects on the macroscopic behavior of water. For example, water is able to dissolve ionic compounds because of its large dipole moment. The anomalously high boiling point of water is also a consequence of its extreme polarity.

Workshop: The Structure of Molecules

1. Molecular models are useful for visualizing molecular shapes and for thinking about molecular properties. It is also quite common to use computer graphics to visualize molecules, particularly large and complex molecules such as proteins, for which building a physical model would be difficult.

In this workshop, you will use a molecular model kit to help visualize simple molecules. The kit consists of colored centers that correspond to various common geometries and connecting joints to link the centers. The connecting joints come in longer and shorter lengths to represent longer and shorter bonds. Use the shorter lengths to indicate bonds to hydrogen and the longer lengths to indicate bonds between any other two atoms.

The colored centers usually follow a standard correlation with atoms:

Black	C	Red	O	White	H
Blue	N	Purple	P	Yellow	S
Silver	metals	Green	halogens		

Each person in the group should choose and complete a row in the table that follows by (a) finding an example of a molecule or an ion with the given structure, (b) predicting the molecular or ionic geometry, (c) building a model, and (d) estimating the bond angles. Share answers with the rest of the group and be prepared to explain them.

A represents a central atom, B represents a terminal atom, and E represents an unshared electron pair on the central atom. We have completed the first line as an example:

Structure	Example	Molecular Geometry	Bond Angles
AB_2	CO_2	linear	180°
AB_2E			
AB_3			
AB_4			
AB_3E			
AB_2E_2			
AB_5			
AB_6			

Answer Questions 2 to 5 in subgroups of two. Share your results and compare models with the rest of the group.

2. a. Build models of each of the following: H_2O, H_2S, CO_2.

   b. Consider the electronegativities of the elements and the three-dimensional molecular geometry of each of the three molecules, and match each of the following dipole moments to its corresponding molecule: 0 D, 0.95 D, 1.85 D. Indicate the electronegative and electropositive regions of each molecule.

3. a. Draw a Lewis diagram of acetic acid (CH_3COOH).

   b. Build a model of the acetic acid molecule.

   c. Identify the electron-pair and molecular geometry around each carbon atom and oxygen atom in the molecule.

   d. Does the molecule have only one unique shape? What factors contribute to its shape or shapes?

4. Benzene (C_6H_6) is an example of a resonance-stabilized molecule. All of the carbon atoms in benzene are connected to each other in a six-atom ring and have a trigonal planar molecular geometry. In contrast, cyclohexane (C_6H_{12}, also a six-carbon ring molecule) is not resonance stabilized. All of its carbon atoms have a tetrahedral geometry.

   a. Draw a Lewis diagram of each molecule.

   b. Build a molecular model of each molecule.

   c. Sketch a three-dimensional representation of each molecule.

   d. Cyclohexane exists in two different shapes known as the boat and chair conformations. These names are descriptive of the shapes. Flex or twist the ends of your model to create each shape, and discuss the relative stability of each. Which is the most stable form? Why?

5. Three molecules known as "substituted benzenes," in which a benzene hydrogen atom is substituted with another atom are (a) C_6H_5Br, (b) C_6H_5F, and (c) $C_6H_4F_2$ (consider only the molecule in which the fluorines are 180° from one another).

   a. Build a molecular model of each molecule.

   b. Predict the order of magnitude of the dipole moments of the molecules, from smallest to largest.

6. The Dutch chemist Jacobus Van't Hoff (1852–1911) and the French chemist Joseph Le Bel (1847–1930) reasoned that carbon, with four covalent bonds, must have a tetrahedral geometry. Let's explore some experimental data that were available to them, as well as some information that was known at the time of their insight, to discover how they reached their conclusion. Use your molecular model kit to help with this task.

Some molecules have the same empirical or molecular formula, but can be separated into compounds that exhibit different physical and chemical properties because they have different molecular structures. These compounds are called *isomers.*

A common instrument used to examine compounds in the late 19th century was a device known as a *polarimeter.* It worked by passing a beam of polarized light through a solution to an analyzing filter. Isomers that are mirror images of each other (called *optical isomers*) rotate the light of the polarimeter in opposite directions.

Van't Hoff examined a number of carbon compounds with the polarimeter. Some had only one isomer, while others had more than one. Consider the following data:

Compound	**Number of Isomers**
CH_4	1
CH_3Cl	1
CH_2Cl_2	1
CH_2ClBr	1
CHClBrF	2

Assume that carbon compounds can be either tetrahedral or square planar. Draw as many isomers as you can, using each of these two geometries. Show how a square planar geometry is inconsistent with the data given, but a tetrahedral geometry logically follows from the data.

7. The compound CHClBrF has two optical isomers.

Stereoisomers are isomers with the same connectivity, or order of attachment among the atoms, but a different orientation of their atoms in space. The stereoisomer classification is further subdivided into enantiomers and diastereomers. *Enantiomers* are stereoisomers whose molecules are nonsuperimposable mirror images of each other, and *diastereomers* are stereoisomers whose molecules are not mirror images of each other.

a. Build molecular models of the two enantiomers of CHClBrF.

b. Consider your hands. Are they mirror images of each other? You may want to use a mirror to check. Are the two enantiomers of CHClBrF mirror images of each other?

c. Sketch three-dimensional representations of the CHClBrF isomers to support your answer in part b.

8. The electron configuration of a carbon atom in its ground state is $1s^22s^22p^2$. This tells us that the 1*s* and the 2*s* orbitals are filled and that two of the 2*p* orbitals have one electron each, with one empty 2*p* orbital. If covalent bonds occur by the overlap of half-filled orbitals, we can assume that carbon should form two bonds. Yet it forms four equivalent single bonds.

   a. What bond angles would result if CH_2 were formed from bonds to the two half-filled *p* orbitals? What is the bond angle in CH_4?

   b. What model do chemists use to explain the fact that the stable carbon–hydrogen compound that forms is CH_4? Explain.

   c. What orbitals are used for bonding by the boron atom in boron trifluoride? Explain.

   d. What orbitals are used for bonding by the beryllium atom in beryllium hydride? Explain.

9. Following are the general energy-level diagrams for second-period homonuclear diatomic-molecule molecular orbitals:

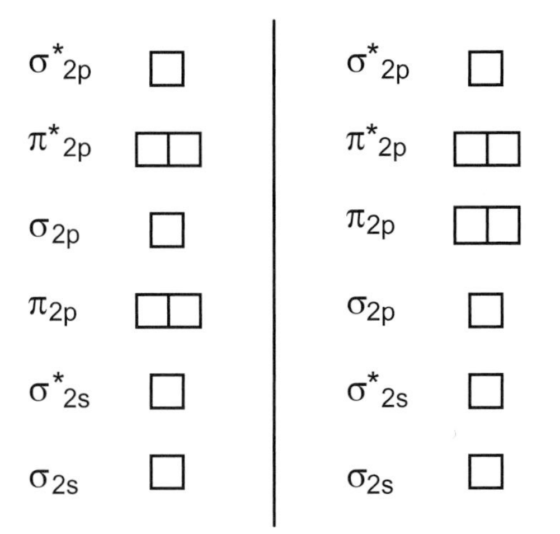

The diagram on the left is for B_2, C_2, and N_2. The diagram on the right is for O_2, F_2, and Ne_2.

a. Construct the molecular orbital electron configuration for each of the six diatomic molecules given.

b. Determine the bond order of each molecule.

c. Determine whether each molecule is paramagnetic or diamagnetic.

Oxidation–Reduction Reactions

> *"The most striking feature of this process is the ease with which a doubly or trebly charged ion gives up a third or fourth electron. We know that this process takes place at the enormously high temperatures in the center of a star, but it is at first a little surprising to find it happening so readily at room temperature."*
> *Ronald W. Gurney*

IN 1884, Svante Arrhenius, a young Swedish scientist, published his doctoral dissertation, in which he described his measurements on the electrical conductivity of salt solutions. In the dissertation, he presented ideas on why the electrical conductivity of water increases greatly when a salt is dissolved in it. Three years later, he published a detailed paper in which he explained his theory of ionic dissociation. Arrhenius assumed that a solution of sodium chloride contained sodium and chloride ions. When electrodes are connected to a battery and placed into the solution, the sodium ions are attracted to the cathode (the negative electrode) and move in that direction. The chloride ions are attracted to the anode (the positive electrode) and move in that direction. This motion of ions through the solution provides the mechanism by which the solution conducts an electrical current.

Arrhenius and other scientists also noticed that the electrolysis of dilute aqueous solutions of electrolytes produced oxygen and hydrogen gases at the electrodes. Hydrogen gas was produced at the cathode and oxygen gas was produced at the anode. They eventually concluded that the chemical reaction at the cathode involved a gain of electrons whereas the reaction at the anode involved loss of electrons.

Chemical reactions in which there is a transfer of electrons from one substance to another are known as *oxidation–reduction reactions,* or *redox reactions.* In this unit, you will examine the oxidation–reduction process and use the concepts of oxidation state and oxidation number to identify redox reactions and to keep track of electrons transferred in a reaction.

Oxidation State

Formal charge is a measure of the charge accumulation on an atom in a bonding situation. The formal-charge concept is useful in describing the electron distribution in molecular compounds. The concept is also helpful in describing resonance structures of molecules and ions. In dealing with redox reactions, however, formal charge is not as useful, because there is a substantial shift in electron density as electrons are transferred from one substance to another. We therefore use the concept of oxidation state to keep track of the location of electrons in redox reactions.

The *oxidation state* is defined as the charge on an atom in a bonding situation if all of the bonded electrons go to the most electronegative atom. The Lewis structure of a molecule or an ion is helpful in determining the oxidation state of an atom in a bonding situation. The oxidation state of the atom is represented by a positive or negative number called its *oxidation number.* Chemists use the following rules to determine the oxidation state of an atom or ion:

1. A free atom has an oxidation number of zero. It is not sharing, gaining, or losing electrons.

2. Polyatomic elements have an oxidation number of zero for each atom. Elements such as H_2, O_2, and P_4 share electrons equally among all atoms in the molecule.

3. The sum of the oxidation numbers in a neutral molecule is zero.

4. A monatomic ion has an oxidation number equal to its charge. For example, the oxidation number of the oxygen in the oxide ion, O^{2-}, is –2.

5. The sum of the oxidation numbers in a polyatomic ion is equal to the charge on the ion.

Let's examine the oxidation numbers of some common elements. Notice the periodic trend among the main-group elements.

Hydrogen Hydrogen has an electronegativity slightly below that of carbon, so when it forms compounds with nonmetals, it usually has an oxidation state of +1. When hydrogen is bonded with metals, it has an oxidation state of –1, because most metals have electronegativity values less than that of hydrogen. Compounds in which hydrogen is in the –1 oxidation state are called *hydrides.*

Group 1A Elements Alkali metals have very low electronegativity values. When they combine with other atoms, they always lose an electron, which gives them an oxidation state of +1.

Group 2A Elements Alkaline earth metals have low electronegativity values. When they form compounds, they give up their two valence electrons, leaving the ions with an oxidation state of +2. A typical example is magnesium oxide (MgO) in which the oxidation state of magnesium is +2.

Aluminum The Group 3A elements have variable oxidation states, but aluminum almost always has an oxidation number of +3 and boron commonly has a +3 oxidation state.

Carbon The concept of oxidation state is rarely applied to carbon compounds, because they typically have a great variety of single, double, and triple bonds and combinations thereof. Oxidation numbers are applied, however, in compounds in which carbon is combined with only one other element. When carbon combines with an element that has a greater electronegativity than itself, it will lose its electrons and have a positive oxidation state. For example, carbon has a +2 oxidation state in CO and a +4 oxidation state in CO_2. When carbon combines with less electronegative elements, it will gain electrons. Carbon has an oxidation state of –4 in CH_4.

Nitrogen The electronegativity value for nitrogen is 3.0, and its oxidation state can vary widely, depending on the atom it combines with. When nitrogen combines with a less electronegative element, such as hydrogen or a metal, its oxidation state is –3, as is the case with NH_3. When nitrogen combines with a more electronegative element, its oxidation state will be positive, varying from +2, as in NO, to +5, as in HNO_3.

Oxygen The term "oxidation" originally meant "combination with oxygen to form an oxide." With an electronegativity of 3.5, oxygen almost always gains electrons when it forms a compound. Molecular oxygen compounds normally have two covalent bonds, and the bonding electrons are strongly held by the oxygen atom. Therefore, oxygen has an oxidation state of –2 in most of its compounds.

Fluorine With the exception of the noble gases, fluorine has the highest electronegativity of all elements, with a value of 4.0; therefore, it always gains an electron and has an oxidation state of –1. The other halogens frequently have an oxidation state of –1 as well, but since electronegativity decreases down a group in the periodic table, they can have multiple oxidation states.

Self-Test 1

1. Identify the oxidation number of each element in the following compounds and ions:

   a. Na_2SO_4

   b. MnO_4^-

   c. $KClO_3$

   d. CH_3OH

2. Identify the oxidation number of nitrogen in the following molecules and ions:

a. NO_3^- d. NO_2 g. NO_2^-

b. HNO_2 e. NO h. N_2O

c. N_2 f. NH_2OH i. N_2H_4

The Transition Metals

The elements in the *d*- and *f*-blocks of the periodic table are called the *transition elements.* Since they are all metals, they are also referred to as *transition metals.* Within this group of metals are two inner groups called the *lanthanide* and *actinide* series.

Many transition metal compounds do not obey the octet rule, and the metals therefore have variable oxidation states. Manganese, for example, has oxidation states ranging from +2 to +7, and iron commonly has oxidation states of +2 and +3. The oxidation state of a transition metal can usually be determined from a knowledge of the oxidation states of the other elements in the compound. In CuO, copper has an oxidation state of +2, because oxygen is –2 and the sum of the oxidation numbers must be zero. In $KMnO_4$, potassium has an oxidation state of +1, oxygen –2; therefore, $(+1) + ? + [4 \times (-2)] = 0$, and the oxidation state of manganese must be +7.

Self-Test 2

1. Determine the oxidation number of the specified element in the following compounds and ions:

a. Cr in $Cr_2O_7^{2-}$ c. Mn in MnO_2

b. Fe in Fe_2O_3 d. Ag in AgCl

Oxidizing and Reducing Agents

Chemists often use the terms "oxidizing agent" and "reducing agent" to describe the reactants in redox reactions. Since oxidation involves the loss of electrons, an oxidizing agent is any substance that can cause a loss of electrons in another substance in a chemical reaction. The oxidizing agent gains electrons, and its oxidation number decreases. Reduction, by contrast, involves a gain of electrons; therefore, a reducing agent is a substance than can cause another substance to gain electrons. The reducing agent loses electrons, and its oxidation number increases.

Consider the reaction of iron(III) oxide and carbon monoxide:

$$Fe_2O_3 + 3\,CO \rightarrow 2\,Fe + 3\,CO_2$$

1. The oxidation number of oxygen is –2 in all of that element's compounds; therefore, oxygen is neither oxidized nor reduced.

2. The oxidation number of iron decreases from +3 in Fe_2O_3 to zero in the uncombined element Fe. Therefore, iron gains electrons, and Fe_2O_3 is the oxidizing agent.

3. Iron is reduced.

4. The oxidation state of carbon increases from +2 in CO to +4 in CO_2. Therefore, carbon loses electrons, and CO is the reducing agent.

5. Carbon is oxidized.

Self-Test 3

1. Consider the reaction $C(g) + O_2(g) \rightarrow CO_2(g)$.

   a. Draw the Lewis structure of CO_2.

   b. Use the Lewis structure to determine the oxidation numbers of carbon and oxygen in CO_2.

   c. What is the oxidation number of the reactants carbon and oxygen?

   d. Which substance is the oxidizing agent?

   e. Which substance is the reducing agent?

Workshop: Oxidation–Reduction Reactions

Questions 1–9: Consider the following activity series:

Metals that react with water to form hydrogen gas and a metal hydroxide:	K Ca Na
Metals that react with steam to form hydrogen gas and a metal hydroxide:	Mg Al Zn Cr Fe
Metals that react with acid to form hydrogen gas and a metal ion:	Ni Sn Pb
Hydrogen:	H_2
Metals that do not react with acids:	Cu Hg Ag Pt Au

This series is used to predict ionic displacement reaction products. The series lists the ability of metals to displace hydrogen from water and acids.

A group round-robin approach has been found to work well for this set of questions.

1. Consider the descriptions of the reactions for the first three groups of metals listed in the activity series. Choose one element from each group and write the reaction that identifies the group.

2. Which metal is the most active? Which is the least active? Explain.

3. List the electronegativity of each of the elements in the activity series. (Refer to Unit 10 for electronegativity values.) What is the relationship between electronegativity and activity?

4. Use the activity series or electronegativity values (or both) to predict whether each of the reactions that follow will occur. If a reaction occurs, write the formulas of the reactants and products and balance the equation.

   a. Zinc and silver nitrate

   b. Gold and sodium chloride

5. Write the equation for the reactions (if any) of the metals sodium, zinc, and gold with hydrochloric acid. Show the changes in oxidation number for each element.

6. Draw the Lewis diagram of H_2O_2. Why do the oxygen atoms have an oxidation state of –1?

7. Compare the oxidation state of hydrogen in HCl with its oxidation state in NaH. Why are they different?

8. Describe the trend in oxidation number for main-group elements as you move from left to right across a row of the periodic table.

9. Consider the reaction $2\ Fe(NO_3)_3(aq) + 3\ H_2S(aq) \rightarrow 2\ FeS(s) + 6\ HNO_3(aq) + S(s)$. Identify the oxidizing agent, reducing agent, substance oxidized, and substance reduced.

Questions 10–14: Balancing redox reaction equations is a skill that combines knowledge of chemistry, common sense, and intuition. There are many methods for balancing redox reactions, and if your instructor has expressed a preferred method, we advise that you follow it, ignoring the remainder of our discussion. Use your instructor's method to complete Questions 10–14.

Otherwise, consider "Sophie's method" for balancing redox equations. Sophie was a workshop leader in the early developmental days of Peer-Led Team Learning. (The authors wish to express their gratitude to Sophie for her insight and contributions to this unit.) Her method relates the number of electrons transferred to the change in oxidation number and consists of the following steps:

1. Identify the changes in oxidation states, and write the oxidation half-reaction and the reduction half-reaction.

For each half-reaction,

2. Balance the atom undergoing redox changes, if necessary.

3. Add the number of electrons that correspond to the change in oxidation state. For reduction half-reactions, add the electrons to the left side of the equation; for oxidation half-reactions, add the electrons to the right side of the equation.

4. Balance the charge in each half-reaction by adding H^+ in acidic solution or adding OH^- in basic solution.

5. Add H_2O to balance oxygen and hydrogen.

6. Multiply the half-reactions by the factor needed to equalize the number of electrons. Add the half-reactions to achieve a final completed overall reaction.

As an example, balance the following redox reaction, which occurs in acidic solution:

$$Fe^{2+}(aq) + MnO_4^-(aq) \rightarrow Fe^{3+}(aq) + Mn^{2+}(aq)$$

Solution:

Step 1:

$$\overset{+2}{Fe^{2+}}(aq) + \overset{+7}{MnO_4^-}(aq) \rightarrow \overset{+3}{Fe^{3+}}(aq) + \overset{+2}{Mn^{2+}}(aq)$$

$Fe^{2+}(aq) \rightarrow Fe^{3+}(aq)$ (oxidation half-reaction)

$MnO_4^-(aq) \rightarrow Mn^{2+}(aq)$ (reduction half-reaction)

Step 2: 1 Fe on each side; 1 Mn on each side; no adjustment necessary

Step 3: $Fe^{2+}(aq) \rightarrow Fe^{3+}(aq) + e^-$ (oxidation half-reaction)

$5\,e^- + MnO_4^-(aq) \rightarrow Mn^{2+}(aq)$ (reduction half-reaction)

Step 4: $Fe^{2+}(aq) \rightarrow Fe^{3+}(aq) + e^-$ (2+ on each side; no H^+ necessary)

$8\,H^+(aq) + 5\,e^- + MnO_4^-(aq) \rightarrow Mn^{2+}(aq)$

Step 5: $Fe^{2+}(aq) \rightarrow Fe^{3+}(aq) + e^-$ (no H_2O necessary)

$8\,H^+(aq) + 5\,e^- + MnO_4^-(aq) \rightarrow Mn^{2+}(aq) + 4\,H_2O(\ell)$

Step 6: $5 \times [Fe^{2+}(aq) \rightarrow Fe^{3+}(aq) + e^-]$

$8\,H^+(aq) + 5\,e^- + MnO_4^-(aq) \rightarrow Mn^{2+}(aq) + 4\,H_2O(\ell)$

$5\,Fe^{2+}(aq) + 8\,H^+(aq) + MnO_4^-(aq) \rightarrow 5\,Fe^{3+}(aq) + Mn^{2+}(aq) + 4\,H_2O(\ell)$

Balance the following equations in acidic solution:

10. $Cu(s) + NO_3^-(aq) \rightarrow NO(g) + Cu^{2+}(aq)$

11. $Cu_2Cl_2(s) + HClO(aq) \rightarrow Cl^-(aq) + Cu^{2+}(aq)$

Balance the following equations in basic solution:

12. $Ca(s) + H_2O(\ell) \rightarrow Ca(OH)_2(s) + H_2(g)$

13. $Fe(OH)_3(s) + OCl^-(aq) \rightarrow FeO_4^{2-}(aq) + Cl^-(aq)$

14. A *disproportionation* reaction is a reaction in which a species is both oxidized and reduced. Consider the self-reaction of nitrogen dioxide in acidic solution to form nitrogen compounds of both lower and higher oxidation states:

$NO_2(g) \rightarrow NO_3^-(aq) + NO(g)$

Balance this redox equation.

Questions 15–19: If time permits, balance the additional practice problems that follow. Otherwise, complete these problems as a postworkshop exercise.

Balance the following equations in acidic solution:

15. $P_4(s) + NO_3^-(aq) \rightarrow H_2PO_4^-(aq) + NO(g)$

16. $Cr_2O_7^{2-}(aq) + Fe^{2+}(aq) \rightarrow Fe^{3+}(aq) + Cr^{3+}(aq)$

17. $HAsO_2^{2-}(aq) + H_2O_2(aq) \rightarrow H_3AsO_4(aq) + H_2O(\ell)$

Balance the following equations in basic solution:

18. $CrO(s) + ClO^-(aq) \rightarrow CrO_4^{2-}(aq) + Cl^-(aq)$

19. $ClO_2(aq) + OH^-(aq) \rightarrow ClO_3^-(aq) + Cl^-(aq) + H_2O(\ell)$

Solutions

> *"It is clear that under these circumstances the classical theory can not be retained. All experimental material indicates that its fundamental starting point should be abandoned, and that, in particular, an equilibrium calculated on the basis of the mass action law does not correspond to the actual phenomena."*
> *P. Debye and E. Hückel*

To a chemist, a solution is nothing more than a homogeneous mixture. The defining phrase, or *definiens,* is composed of the two words "homogeneous" and "mixture." Each word has a very specific meaning in chemistry: *Homogeneous* means that a sample has a uniform appearance and composition throughout. *Mixture* denotes a sample that consists of two or more substances. If both of these definitions are met, a sample is a solution.

You encounter solutions frequently, both in the chemistry laboratory and in everyday life. The air we breathe and the oceans, lakes, and streams that cover most of our planet are examples of solutions. In the laboratory, solutions are an excellent medium for the promotion of chemical reactions. In a solution, the particles are much closer together than in a gas, and they have more freedom of movement than in a solid. Outside of the laboratory, the process of life itself depends on solutions.

Solution Terminology

A specialized vocabulary is used in discussing solutions. Let's look at some of these terms and see how they are applied.

In general, the component of the solution that is present in the greatest amount is called the *solvent.* The substance with the smaller amount is called the *solute.* These terms are not precise, however, and their usage varies among different specialties in chemistry. For example, in water solutions, water is almost always referred to as the solvent, no matter its relative amount. Also, when a solid or a gas is dissolved in a liquid, the liquid is generally called the solvent.

It is often useful to know the maximum amount of a solute that will dissolve in a given solvent at a specified temperature. This measure is known as the *solubility* of that solute. Reference sources often report solubilities in grams of solute per 100 grams of solvent. When a solution contains an amount of solute less than the solubility limit, it is said to be *unsaturated*; if it is at the solubility limit, it is *saturated*. Under certain special conditions, a solution can contain more solute than its normal solubility limit, and in this case it is called *supersaturated*.

The terms "concentrated" and "dilute" are often used to describe solutions. It is important to keep in mind that these terms are valid only in a relative sense. A *concentrated* solution has a relatively large amount of solute per given amount of solvent compared with a *dilute* solution. The comparison is valid only for systems with the same solute and solvent.

In discussing solutions of liquids in liquids, chemists use the term *miscible* to describe two liquids that will dissolve in one another in all possible combinations. When liquids will not dissolve in one another, they are said to be *immiscible*. A chemist would say that oil and water are immiscible, whereas alcohol and water are miscible.

Self-Test 1

1. Solution A is made by dissolving 0.250 g of an ionic salt in 1.00 L of water, and solution B is made by dissolving 2.50 g of a different ionic salt in 0.100 L of water. Which solution is dilute and which is concentrated? What are the solute and solvent in each solution? Which solution is saturated and which is unsaturated? Are these salts miscible or immiscible in water? Can all of these questions be answered?

Solution Concentration Units

A number of different units are used to express the quantity of solute dissolved in a given amount of solvent. Common units include percentage by mass, parts per million, molarity, and molality. The definition of each provides the basis for calculations with that unit:

$$\% \text{ by mass} = \frac{\text{mass solute}}{\text{mass solution}} \times 100 \qquad \textbf{(13.1)}$$

$$\text{parts per million} = \text{ppm} = \frac{\text{mass solute}}{\text{mass solution}} \times 10^6 \qquad \textbf{(13.2)}$$

$$\text{molarity} = \text{M} = \frac{\text{moles solute}}{\text{liters solution}} \qquad \textbf{(13.3)}$$

$$\text{molality} = \text{m} = \frac{\text{moles solute}}{\text{kilograms solvent}} \qquad \textbf{(13.4)}$$

The choice of concentration unit is largely a matter of convenience. There are some technical considerations that must be made, however: Percentage by mass, parts per million, and molality are applicable at any temperature; molarity, in contrast, is dependent on temperature, because the volume of the solution, measured in liters, varies with temperature.

Self-Test 2

1. A solution is made by dissolving 1.00 g of sodium chloride in 1.00 L of water. Assume that the volume of the resulting solution is 1.00 L and that the density of both water and the resulting solution is 1.00 g/mL. Determine the concentration of the solution as percentage by mass, parts per million, molarity, and molality.

Formation of Solutions

Two processes contribute to the formation of a solution: (1) a change in the heat energy content of the system and (2) a change in the disorder of the system. The sum of the effects of these changes must result in an overall release of energy available to do work if a solution is to form. Let's consider each process separately.

The changes in heat energy that result in the formation of a solution consist of interactions (a) among the solute particles, (b) among the solvent particles, and (c) between the solute and solvent particles. Consider the process by which table salt—sodium chloride—dissolves in water. The positively charged sodium ions and the negatively charged chloride ions are attracted to one another. When a solution forms, these ions must be separated from each other. Energy is required to separate the solute particles. Similarly, energy is needed to separate the solvent particles. The water molecules are clumped together in hydrogen-bonded groups that must be "pulled apart" so that solute ions can fit between them. This process also requires an input of energy. The negative ends of water molecules then surround positively charged sodium ions, and the positive ends of water molecules surround negatively charged chloride ions. Energy is released when the solute and solvent particles interact.

The other process to consider in analyzing the formation of a solution is the tendency in nature toward increasing disorder. In most cases, when solutions form, there is an increase in disorder that results from the separation of the solute and solvent particles. This is an energetically favorable process.

Colligative Properties of Solutions

Solutions are unique in that some of their properties depend only on the concentration of solute particles, without regard to their identity. These are called *colligative properties*. The colligative property concept has important limitations, however, chiefly that it applies only to the *change* in the properties of *dilute* solutions.

Freezing-point depression and boiling-point elevation

Probably the most commonly utilized application of colligative properties is the solution of ethylene glycol and water found in automobile radiators. The ethylene glycol solute is usually labeled as "winter antifreeze and summer coolant." Its effect is to both lower the freezing temperature and raise the boiling temperature compared with these properties of pure water alone. Both freezing-point depression and boiling-point elevation are colligative properties.

Freezing-point depression is the change in freezing point that occurs in a solution compared with the freezing point of the pure solvent. It is directly proportional to the molality of the solution:

$$\Delta T_f \propto m \qquad \textbf{(13.5)}$$

If we introduce a proportionality constant, we have the equation

$$\Delta T_f = -K_f \times m \tag{13.6}$$

where K_f is the *molal freezing point depression constant.* This constant is valid for any solute in a dilute solution of the specified solvent. The negative sign accounts for the "depression."

Boiling-point elevation is the change in boiling point of a solution compared with the boiling point of the pure solvent. As with freezing-point depression, boiling-point elevation is proportional to the molality of the solution:

$$\Delta T_b \propto m \tag{13.7}$$

The introduction of a proportionality constant called the *molal boiling-point elevation constant* gives us an equation:

$$\Delta T_b = K_b \times m \tag{13.8}$$

Since water is the most common solvent, the freezing- and boiling-point constants for water are those most frequently encountered: K_f = 1.86°C · kg solvent/mol solute and K_b = 0.52°C · kg solvent/mol solute. A bit of common sense comes in handy here in working with these cumbersome units. As long as calculations involving Equations 13.6 and 13.8 are not mixed with other calculations, we can substitute 1.86°C/m and 0.52°C/m for the freezing-point and boiling-point constants, respectively.

Self-Test 3

1. Determine the freezing and boiling points of a solution made by dissolving 30.0 g of glucose ($C_6H_{12}O_6$) in 100.0 g of water.

Lowering of vapor pressure

Another colligative property of a solution is found in comparing the vapor pressure of a solution with that of the pure solvent. Before discussing this colligative property, however, we must introduce a new concentration unit called the *mole fraction,* which is, by definition

$$\text{mole fraction} = X = \frac{\text{moles solute}}{\text{total moles of solution}} \quad \textbf{(13.9)}$$

The *vapor pressure* of a solution is directly proportional to the mole fraction of the solvent particles:

$$P_{soln} = X_{solvent} \times P_{solvent} \quad \textbf{(13.10)}$$

However, we will be interested in the change in vapor pressure that occurs by the addition of a solute. Subtracting both sides of Equation 13.10 from the identity $P_{solvent} = P_{solvent}$, we have

$$P_{solvent} - P_{soln} = P_{solvent} - (X_{solvent} \times P_{solvent}) \quad \textbf{(13.11)}$$

Notice how the left side of Equation 13.11 is the difference between the vapor pressure of the pure solvent and that of the solution. This is the change in vapor pressure, ΔP, that will occur upon addition of the solute. We can also factor out $P_{solvent}$ from the right side of Equation 13.11 to obtain

$$\Delta P = P_{solvent} \times (1 - X_{solvent}) \quad \textbf{(13.12)}$$

If we add the mole fraction of the solvent plus the mole fraction of the solute, the total must equal unity:

$$X_{solvent} + X_{solute} = 1 \quad \textbf{(13.13)}$$

Solving Equation 13.13 for the mole fraction of the solute yields

$$X_{solute} = 1 - X_{solvent} \quad \textbf{(13.14)}$$

Finally, substituting Equation 13.14 into Equation 13.12 results in

$$\Delta P = P_{solvent} \times X_{solute} \quad \textbf{(13.15)}$$

Thus, we see that the change in the vapor pressure of a solution depends on the mole fraction of *solute* particles: The more solute particles, the greater is the change in vapor pressure.

Osmotic pressure

Osmotic pressure is the final colligative property that we will consider in this unit. Osmotic pressure is the pressure required to prevent the phenomenon known as *osmosis,* or diffusion of solvent through a semipermeable membrane. Consider Figure 13.1, which shows a view of a solution on the right side of a semipermeable membrane and pure solvent on the left side The membrane has pores large enough to allow the passage of water molecules, symbolized by the smaller, light circles, but small enough to prevent the passage of solute molecules, symbolized by the larger, shaded circles. The pressure required on the right side of the apparatus to balance the rate of movement of solvent particles between the two compartments is the osmotic pressure of the solution.

Osmotic pressure, symbolized by π, is directly proportional to the molarity of the solution:

$$\pi \propto M \qquad \textbf{(13.16)}$$

The proportionality constant that changes this relationship into an equation is RT, the product of the ideal gas constant and the absolute temperature:

$$\pi = RT \times M \qquad \textbf{(13.17)}$$

Molarity is moles per liter, which is moles (n) per volume (V), or M = n/V. Substituting this into the preceding equation, we find that osmotic pressure is similar to ideal gas pressure in the ideal gas equation:

$$\pi = RT \times \frac{n}{V} \quad \textit{or} \quad \pi V = nRT \qquad \textbf{(13.18)}$$

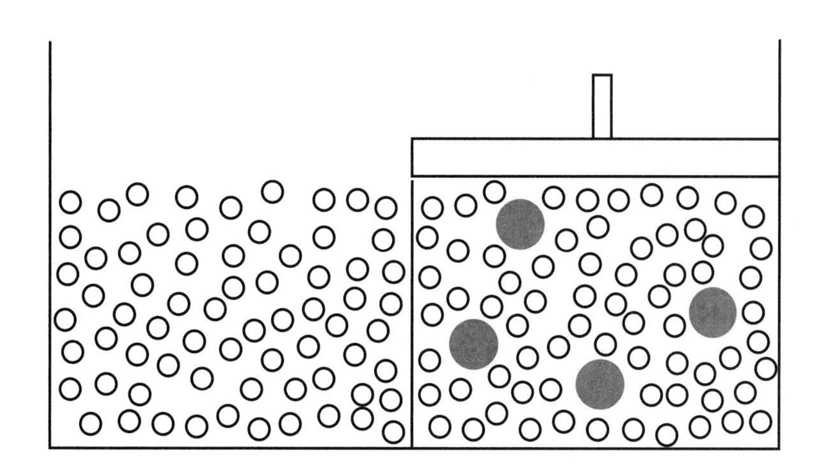

Figure 13.1. Osmotic pressure. The pressure that must be applied to the right side of the apparatus—the side that holds the solution—to prevent net movement of pure solvent from the left side is the osmotic pressure of the solution.

Workshop: Solutions

1. Two liquids are miscible. How can they be separated? Brainstorm to come up with a list of possibilities, and then discuss each item on the list until the group arrives at no fewer than two feasible methods to separate miscible liquids.

2. Split your group into two subgroups. Each subgroup should work on one part of this question. When both subgroups have completed their part of the question, each subgroup should demonstrate its solution and the other subgroup should question and critique the answer. Also, as a complete group, compare and contrast the two parts of the question in order to identify the similarities and differences between the two solution processes.

   Consider the potential formation of two different solutions with the same solvent, namely, water. Solution A has liquid ammonia (NH_3) as the solute, and Solution B has liquid butane ($CH_3CH_2CH_2CH_3$) as the solute. In each case, analyze the relative values of the energy changes involved in (a) separating solute particles, (b) separating solvent particles, and (c) the interaction between solute and solvent particles. Carefully consider the role of intermolecular forces in solute–solvent interaction.

3. Split into the same two subgroups as for Question 2, but for this question work on the solute other than the one considered in the previous question. Use a series of particle-level illustrations or work with physical models such as your molecular model kit to demonstrate what happens when a solution does or does not form. Again, carefully consider the role of intermolecular forces when you plan your demonstration. Each subgroup should make a presentation to the other subgroup when both groups are ready.

4. How can you speed up the process of dissolving a specified amount of solute in a given quantity of solvent? Brainstorm to come up with a list of possibilities, and then narrow your list to no more than three methods. For each of the three methods, analyze what happens at the submicroscopic level to make that method effective in speeding up the dissolving process.

Questions 5 through 12: Use the group round-robin method to answer each question.

5. You need to make 1.00 kg of a 22.5% ammonium chloride solution. How many grams of ammonium chloride do you use? What volume of water will be used? (*Hint:* Recall the density of water.)

6. A sample of fruit juice is analyzed for its lead concentration. If the sample is found to have 0.477 ppm lead, how many grams of lead would be in an 8.0-fl oz serving of the juice? Assume that the density of the juice is 1.01 g/mL.

7. A student is instructed to prepare 3.00 L of a 0.100-M solution of copper(II) sulfate. She goes to the stockroom and finds a 100-g bottle of the anhydrous salt, which cost $14.00, and a 500-mL bottle of a 1-M solution, which cost $12.50. Which source provides the most economical way of making the solution?

8. How would you prepare 5.00 L of a 0.75 molal solution of table sugar ($C_{12}H_{22}O_{11}$)?

9. You are asked to develop a protocol for preventing an automobile radiator from freezing down to –10°F. How many kilograms of ethylene glycol ($HOCH_2CH_2OH$) are needed in 3.0 gallons of water to prevent freezing at any temperature at or above that specified?

10. Determine the molal boiling-point elevation constant for a solvent if the boiling point of a solution made by dissolving 0.75 g of urea (NH_2CONH_2) in 14.0 g of the solvent results in an increase in the boiling point of 2.44°C greater than the boiling point of the pure solvent.

11. A solution is prepared by dissolving 10.0 g of sucrose (common table sugar) in water and diluting the combination to a final volume of 1.00 L. What osmotic pressure would this solution develop at 22°C?

12. A solution is prepared by dissolving 7.33 g of a protein in water at 22°C to make a total of 500 mL of solution. This solution is then analyzed for its osmotic pressure, which is found to be 11.3 mm Hg. What is the molar mass of the protein?

Chemical Kinetics: Concepts and Models

> *"Science is the tool of the Western mind and with it more doors can be opened than with bare hands."*
> *C. G. Jung*

You have witnessed countless chemical reactions throughout your life, and undoubtedly you have observed that some, such as wood burning in a fireplace, occur very quickly, and others, such as rusting of the iron body of a car, occur very slowly. *Chemical kinetics* is the study of the rates of chemical reactions and how the reactions take place.

Perhaps you have a bottle of hydrogen peroxide in your bathroom cabinet. It is usually brown, and its label will have the instructions "store in a cool, dark place." These storage instructions are there because hydrogen peroxide undergoes a decomposition reaction, forming water and oxygen. Fortunately, the reaction is slow enough that your bottle will last for a few years. If you were to sprinkle in some solid manganese dioxide, however, the contents of the bottle would decompose in a matter of seconds. Remarkably, *all* of the manganese dioxide could be filtered off from the remaining water and reused to speed the decomposition of another bottle of hydrogen peroxide. Substances such as manganese dioxide, which speed up a chemical reaction without being consumed in the reaction, are called *catalysts.*

The process of life itself depends on catalytic molecules known as *enzymes.* Most biochemical reactions rely on enzymes that increase the rate of the reaction by 1,000 to 10^{18} times compared with the rate of the uncatalyzed reaction. An even more remarkable property of enzymes is that they produce reaction products at near-100% yields, allowing all of the reaction energy to be used by the cell, as well as virtually eliminating any undesirable by-products.

No matter the type of chemical change—burning, rusting, decomposition, a biological reaction, or any other transformation—understanding reaction kinetics is essential to comprehending the process.

Average and Instantaneous Rate

Suppose you arrive at school one day, and as you wait for your first class to begin, your friend arrives, sits next to you, and remarks, "It was a slow commute today." You ask, "How fast did you travel?" Your friend looks back at you with a puzzled look in her eyes.

How do you respond to someone who asks about the speed of a trip? If the distance from home to school is ten miles and it takes a half hour to drive to school, your average speed, or *average rate* of travel, is 20 miles per hour (mph). However, along the way you may have gone as fast as 45 mph, but at other times you were stopped. Your speed at any given instant, your *instantaneous rate*, is shown by your speedometer reading at that moment. Clearly, there are two ways of expressing rates. Your choice depends on the type of information that is needed.

The rate of a chemical reaction is defined as the change in concentration of a reactant or product of that reaction per unit time. This change can be expressed as $\frac{\Delta\,[\text{reactant}]}{\Delta t}$ or $\frac{\Delta\,[\text{product}]}{\Delta t}$. Just as with the speed of a trip from home to school, chemical reaction rates can be average or instantaneous. To obtain an instantaneous rate, the reaction needs to be examined over a period small enough to ensure that the reaction has proceeded only slightly.

Self-Test 1

1. Which of the following statements express an instantaneous rate and which express an average rate?
   a. It took 58 minutes to drive 40 miles to work today.
   b. A car traveled 1.00 foot in 0.01 second.
   c. A compound fully decomposed in 0.001 second.
   d. A compound is 3.0% decomposed in 0.05 second.

2. Consider the following graph of reactant concentration versus time:

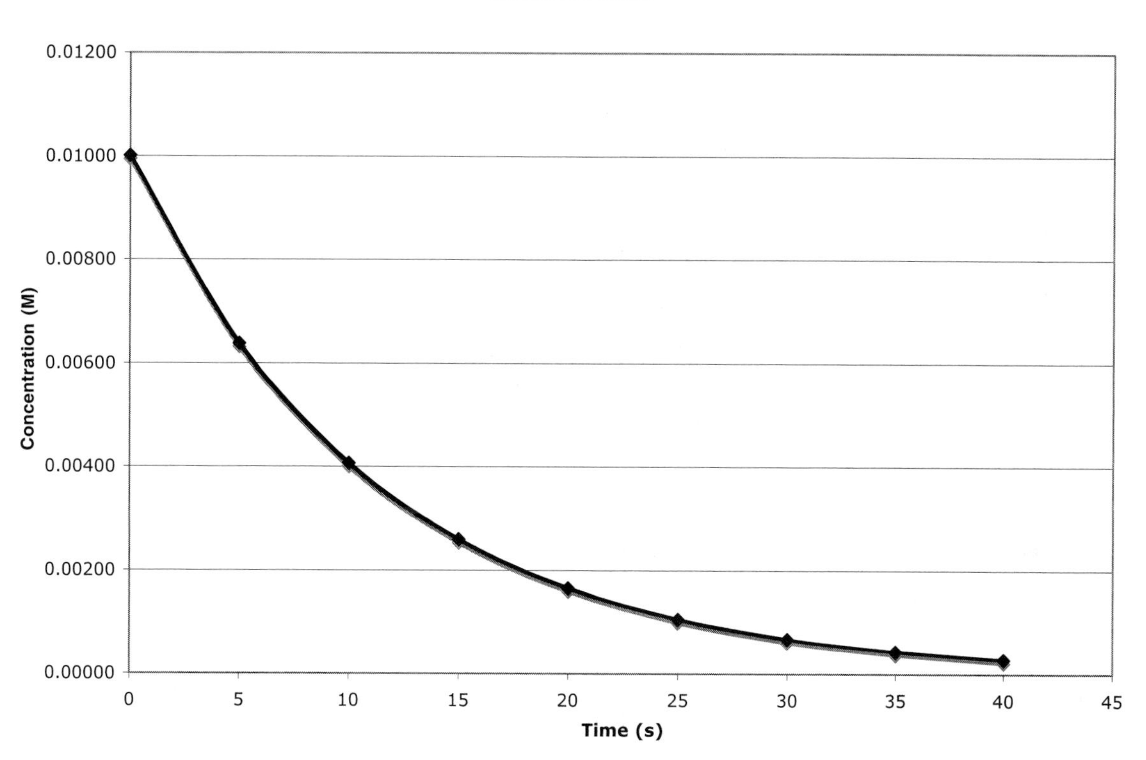

a. What is the rate of reaction between 8 and 12 seconds?

b. What is the rate of reaction over the first 36 seconds?

c. Which of the two rates found in parts (a) and (b) would be regarded as a better estimate of an instantaneous rate? Explain your choice.

Rate Laws

A *rate law* describes how the speed of a reaction depends on the concentrations of reactants and products. A rate law is determined from experiments that measure the change in concentration of one or more of the reactants or products over time. Consider the following rate laws that can apply to the general reaction A + B → Products:

$$\text{rate} = -\frac{\Delta[A]}{\Delta t} = k\,[A] \qquad \textbf{(14.1)}$$

$$\text{rate} = -\frac{\Delta[A]}{\Delta t} = k\,[A]^2 \qquad \textbf{(14.2)}$$

$$\text{rate} = -\frac{\Delta[A]}{\Delta t} = k\,[A]\,[B] \qquad \textbf{(14.3)}$$

$$\text{rate} = -\frac{\Delta[A]}{\Delta t} = k \qquad \textbf{(14.4)}$$

Equations 14.1–14.4 express the rate of the reaction in terms of the disappearance of A: $\text{rate} = -\frac{\Delta[A]}{\Delta t}$. This rate is proportional to the molar concentration of A or B (or both), raised to a power. In each case, the proportionality constant is given by k, which is known as the *rate constant.*

The term *reaction order* is used to refer to the exponents of the concentrations. We can refer to the overall reaction order—the sum of the exponents of all of the reactants—or the reaction order of a particular reactant—the exponent of that reactant. For example, the rate law in Equation 14.2 is said to be second order with respect to A. For Equation 14.3, the reaction is second order overall and first order for each of the reactants. It is important to note that the order of the reaction does not necessarily follow from the reaction stoichiometry.

The rate constant can be expressed in an alternative form known as the *half-life* of the reaction, which is defined as the time needed for the concentration of the reactant to fall to exactly half of its original concentration. Half-lives are most commonly used for first-order reactions. The relationship between the half-life and the rate constant for a first-order reaction is given by

$$t_{1/2} = \frac{\ln 2}{k} = \frac{0.693}{k} \qquad \textbf{(14.5)}$$

Self-Test 2

1. For each of the following rate expressions, state the order of the reaction with respect to B. State the overall reaction order.
   a. rate = k [A] [B]
   b. rate = k [A] $[B]^2$
   c. rate = k

2. A first-order reaction has a rate constant of 100 s^{-1}. Calculate the half-life for this reaction.

3. The half-life for the reaction 2 A(g) $\rightarrow$ 4 B(g) + C(g) is 24.0 min. The initial concentration of A is 1.6×10^{-3} M. What is the concentration of A after 48.0 minutes have elapsed? What is the rate constant?

Reversibility

Often, the rate of a reaction will depend not only on the concentrations of the reactants, but also on the concentrations of the products. For example, the experimentally determined rate law for the reaction

$$A \underset{k_{-1}}{\overset{k_1}{\rightleftharpoons}} B$$

may be

$$\text{rate} = -k_1 [A] + k_{-1} [B]$$

In this case, the reaction is said to be *reversible*. You will see that kinetic reversibility is closely connected to the concept of equilibrium.

Workshop: Chemical Kinetics: Concepts and Models

Questions 1 and 2: Work in pairs or groups of three, as directed by your leader.

1. Consider a simple chemical reaction A → B that follows a first-order rate law rate = k[A]. You will model this reaction with pennies. Start with 100 pennies, which will represent the initial concentration of A, 100 mM. Each penny will therefore represent 1 mM.

   Student A (SA) will represent the concentration of A, and Student B (SB) will represent the concentration of B. We will represent the reaction of A to form B by passing pennies from SA to SB. Each exchange of pennies will represent 1 second of time. SA will do the physical manipulation of the pennies, and SB will record the results. If you are in a group of three, the third person should verify the calculations and record the results.

   We will model a reaction in which 10% of the concentration of A reacts per second. Thus, for each exchange (each second), SA should transfer 10% of the remaining pennies to SB. Round fractions to the nearest penny. Continue this exchange for 15 seconds. Record concentrations of A and B (number of pennies) each second (after each exchange step) in the following table (leave the "ln [A]" column blank for now):

Time (s)	[A] (mM)	[B] (mM)	ln [A]
0	100	0	4.61
1			
2			
3			
4			
5			
6			
7			
8			
9			
10			
11			
12			
13			
14			
15			

Plot the concentration of A versus time on the graph that follows. Use a different color to plot the concentration of B versus time on the same graph.

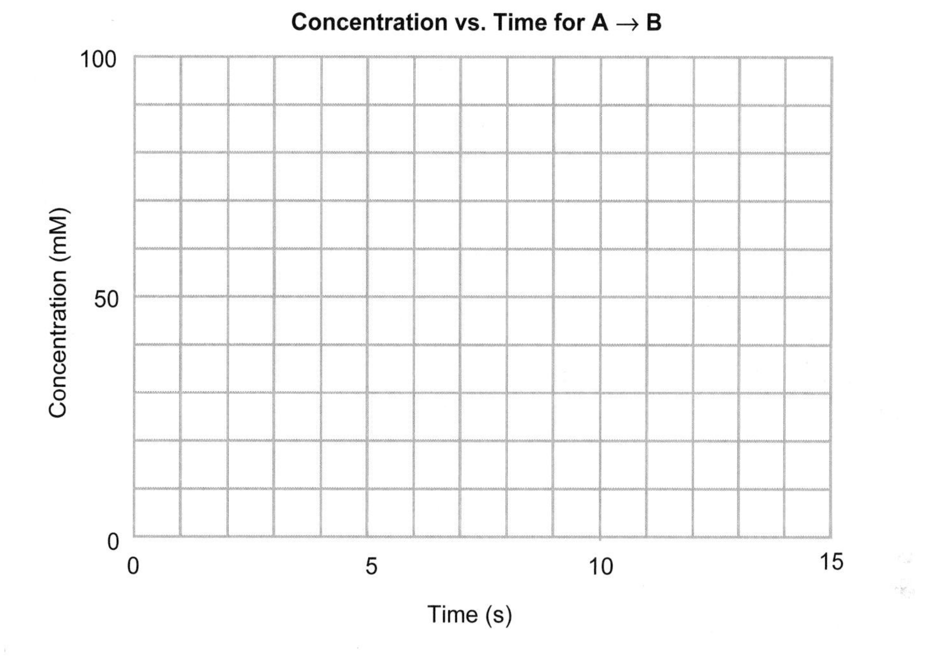

Calculate ln [A], and record the data in the preceding table. Plot ln [A] versus time on the following graph:

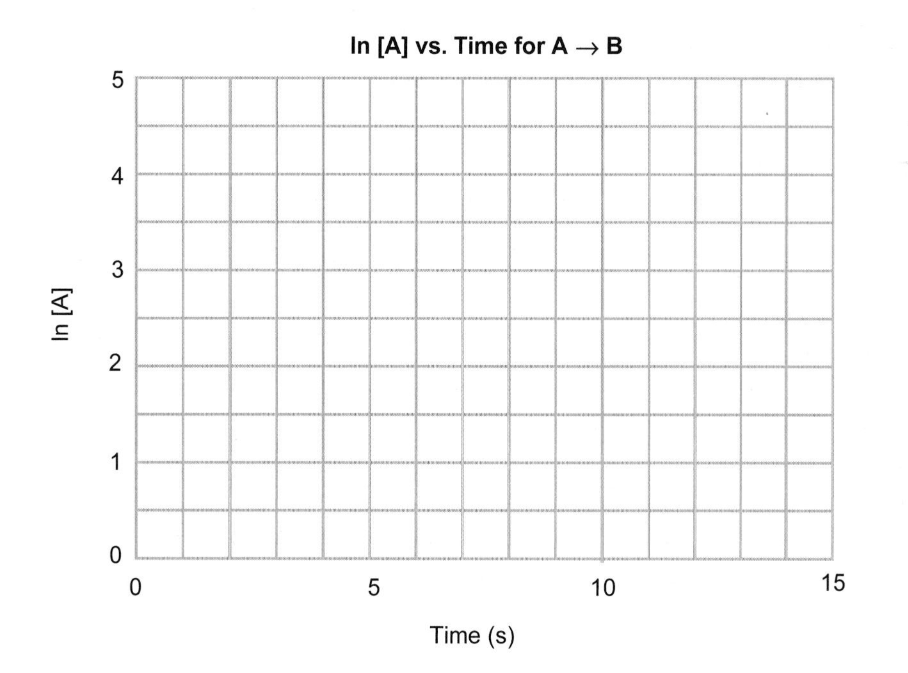

Determine the slope of the best-fit straight line for the plot of ln [A] vs. time. How does this slope relate to the fact that 10% of A reacted per second?

2. Let's apply the modeling technique developed in Question 1 to a reversible reaction, $A \rightleftharpoons B$.

a. In each second (exchange step), allow 10% of A to react to form B, and then allow 10% of B to react to form A. Record the results in the first two columns of the table that follows.

b. In each second (exchange step), allow 10% of A to react to form B, and then allow 5% of B to react to form A. Record the results in the last two columns of the following table:

	10%/10%		10%/5%	
Time (s)	[A] (mM)	[B] (mM)	[A] (mM)	[B] (mM)
0	100	0	100	0
1				
2				
3				
4				
5				
6				
7				
8				
9				
10				
11				
12				
13				
14				
15				

Plot the concentration of A versus time for the 10%/10% reaction on the graph that follows. Use a different color to plot the concentration of B versus time on the same graph.

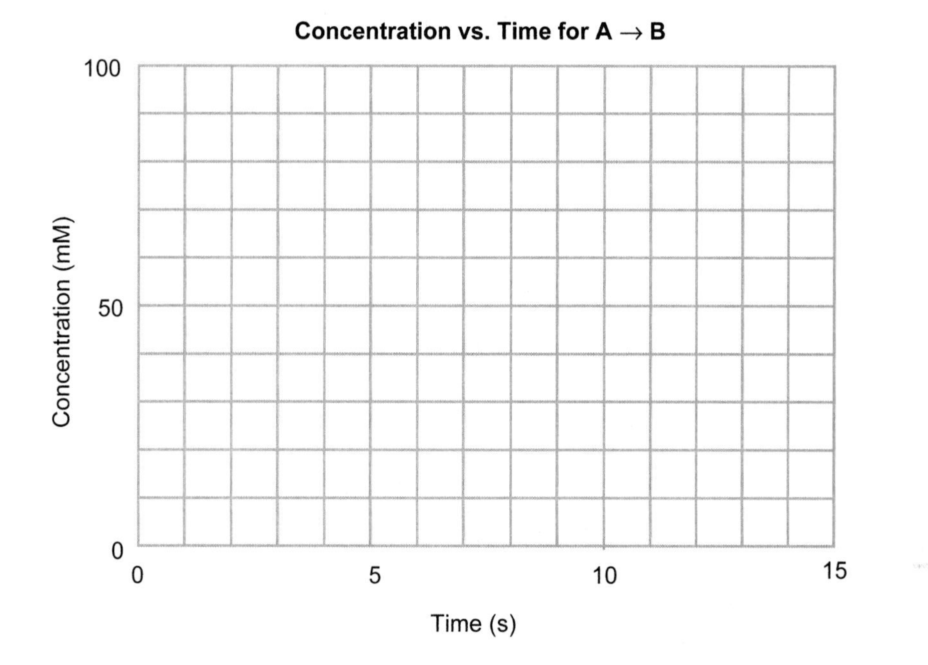

Plot the concentration of A versus time for the 10%/5% reaction on the graph that follows. Use a different color to plot the concentration of B versus time on the same graph.

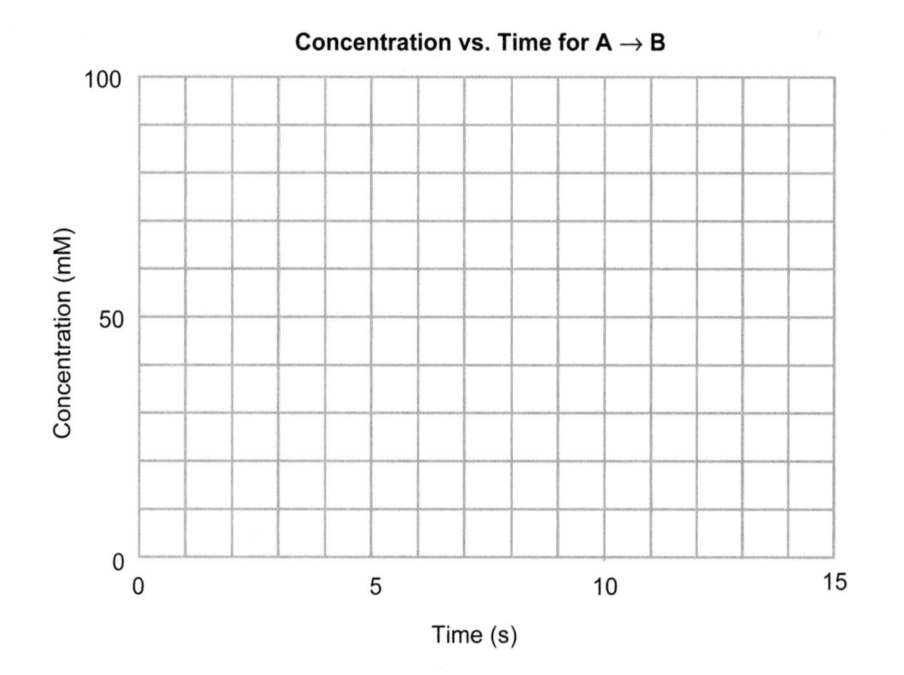

3. The civilization on the planet Ecton is endangered by the continual illegal disposal of Klingon hair spray, which is causing the atmosphere to decompose in accordance with a first-order rate law. Science Officer Spock has determined that their atmosphere is decomposing with a half-life of 12.50 min. Dr. McCoy, the medical officer, has determined that the Ectonians need a minimum of 6.25% of the original atmosphere to survive. Meanwhile, Scottie, the chief engineer, is desperately trying to fix the transporter so that the entire Ectonian population can be transported to a safe planet. You are the acting commander, replacing Captain Kirk, who is on vacation. How long does Scottie have to repair the transporter?

4. What are the units for zeroth-, first-, and second-order rate constants? Use moles, liters, and seconds as the basic units.

5. The following initial rate data were collected for the reaction of hydrogen iodide with ethyl iodide: $HI(g) + C_2H_5I(g) \rightarrow C_2H_6(g) + I_2(g)$

[HI]	$[C_2H_5I]$	Initial Rate (mol/L · s)
0.015	0.900	4.01×10^{-5}
0.030	0.900	8.04×10^{-5}
0.030	0.450	3.99×10^{-5}

From the data, determine the rate law and the value of the rate constant.

6. You are assigned to a team that is conducting research on the decomposition of a protein in the presence of an oxidizing agent in aqueous solution. As the kinetics expert on the team, your job is to determine the rate law for the reaction. You have available a spectrophotometer that can measure the concentration of the protein at any concentration above a minimum of 0.1 mM. Describe the experiments you will perform.

7. Consider again the plot shown in Self-Test 1, Question 2:

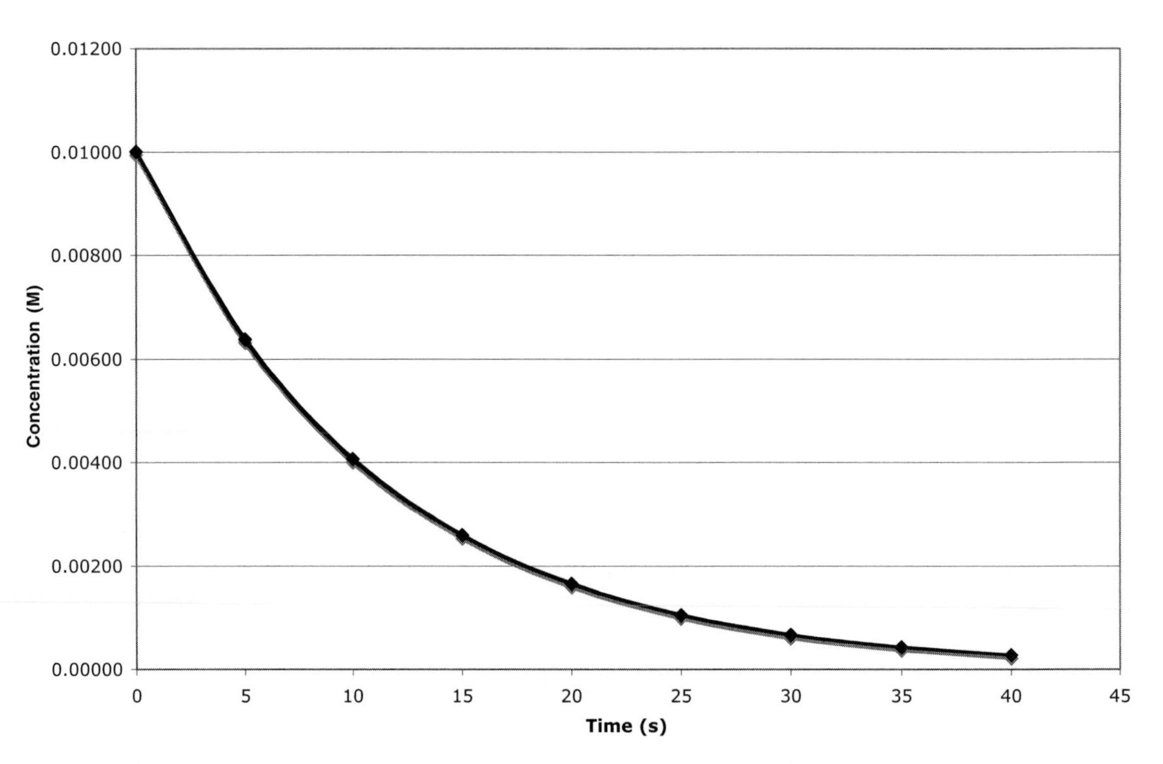

This plot shows a first-order decay with an initial concentration of 0.010 M.

a. Estimate the initial rate.

b. Estimate the rate constant.

c. Estimate the half-life from the rate constant.

d. Is the estimated half-life consistent with the plot?

Rate Data and Rate Equations

> *"Nothing endures but change."*
> *Heraclitus*

If we experimentally measure how the rate of a chemical reaction depends on the concentrations of the reactants and products, we can develop the rate equation for that particular reaction. In this unit, you will learn how to use experimental data to find the rate equation. You will also learn that rates of chemical reactions are affected by concentrations of reactants, temperature, and the presence of a catalyst.

Differential (Instantaneous) Rate Expressions

Many reactions have an instantaneous rate expression that follows the general form

$$\text{rate} = k\,[\text{reactant 1}]^m\,[\text{reactant 2}]^n \ldots$$

where k is the proportionality constant that relates the reaction rate to the concentrations of reactants. The concentration of each species in the reaction is modified by an exponent: m and n in this general case. These exponents denote the *order* of the reaction with respect to each species in the reaction. We would say that the reaction is of mth order in reactant 1, nth order in reactant 2, etc. The sum of the exponents is known as the overall reaction order. Let's look at some common types of differential rate expressions. For the simple reaction $A \rightarrow B$, the rate is defined as the change in concentration of reactant A per unit time.

First-Order Rate Expression

If the reaction is first order, the rate equation is

$$\text{rate} = -\frac{\Delta[A]}{\Delta t} = k\,[A]$$

This equation says that the rate of the reaction is directly proportional to the concentration of A. The proportionality constant, k, is known as the *rate constant*. The concentration unit is molarity, as is implied by the square brackets. A reaction rate is expressed as a positive number. Since the concentration of A is decreasing, $\Delta[A]$ is negative. The minus sign preceding the expression makes the net rate positive. This reaction is first order in A and first order overall.

A first-order rate law essentially says that the speed at which the reaction proceeds is equal to a proportionality constant times the concentration of A. Given that k is a constant, the concentration of A is the only variable that affects the reaction rate. Thus, the reaction rate can be measured by following the change in the concentration of the reactant per change in time. If the concentration of A is doubled, the reaction rate doubles. If the concentration of A is halved, the reaction rate is also halved.

Second-Order Rate Expression

If the reaction is second order, the rate equation is

$$\text{rate} = -\frac{\Delta[A]}{\Delta t} = k\,[A]^2$$

The rate of a second-order reaction is directly proportional to the square of the concentration of A. Here, the reaction is second order with respect to the concentration of A and second order overall. If the concentration of A is doubled, the reaction will proceed four times as fast.

Half-Order Rate Expression

If the reaction is half order, the rate expression is

$$\text{rate} = -\frac{\Delta[A]}{\Delta t} = k\,[A]^{1/2}$$

The rate of a half-order reaction is directly proportional to the square root of the concentration of A. The reaction is half order with respect to A and half order overall. If the concentration of A is doubled, the reaction rate will increase by a factor of $(2)^{1/2}$, or $\sqrt{2}$, which is 1.4.

Zeroth-Order Rate Expression

Reactions in which the rate is independent of the concentrations of the reactants are called zeroth order. The rate expression is

$$\text{rate} = -\frac{\Delta[A]}{\Delta t} = k$$

The rate of a zeroth-order reaction does not depend on the concentration of A.

Self-Test 1

1. State the order with respect to each reactant and the overall order of the reaction for the following rate expressions:

   a. rate = k [A] [B]

   b. rate = k $[A]^2$ [B]

   c. rate = k [A] $[B]^{1/2}$

2. Consider the following rate expression, and then answer the questions that follow:

   The reaction

$$2\,NO(g) + O_2(g) \rightarrow 2\,NO_2(g)$$

   obeys the rate law

$$rate = k\,[NO]^2\,[O_2]$$

   In each case, explain how the rate of the reaction will change when the following changes in concentration are made:

   a. $[O_2]$ is doubled

   b. [NO] is doubled

   c. [NO] is halved

   d. $[O_2]$ is halved and [NO] is doubled

   e. [NO] is halved and $[O_2]$ is doubled

Integrated Rate Expressions

It is frequently important to calculate either the concentration of reactant remaining at a given time or the time required for the concentration of the reactant to reach a certain level. Integrated rate expressions tell us how concentration changes as a function of time. The mathematics involves calculus, and we next state the results for two of the simplest rate laws. Integrated rate expressions are often applied to the problem of determining the order of a chemical reaction.

First Order

$$[A]_t = [A]_{t=0}\, e^{-kt} \quad \textit{or} \quad \ln [A]_t = -kt + \ln [A]_{t=0}$$

From the first-order rate law, you can see that a graph of ln [A] vs. t will result in a straight line with a slope of –k and a y-intercept of $\ln [A]_{t=0}$.

Second Order

$$\frac{1}{[A]_t} = kt + \frac{1}{[A]_{t=0}}$$

From the second-order rate expression, you can see that the graph of $\frac{1}{[A]_t}$ vs. t is a straight line whose slope is k and whose y-intercept is $\frac{1}{[A]_{t=0}}$.

Self-Test 2

1. What is the difference between an instantaneous rate equation and an integrated rate equation? Use graphs or drawings in your explanation.

2. Explain how you can use integrated rate laws to tell whether a reaction is first order or second order. Use graphs and drawings in your explanation.

The Arrhenius Equation

The Swedish chemist Svante Arrhenius was the first to note that the rate constant for many chemical reactions increased with temperature. He formulated a mathematical expression of this relationship that is now known as the Arrhenius equation:

$$k = Ae^{-E_a/RT}$$

Here, A is the frequency (or preexponential) factor and E_a is the activation energy. The frequency factor is the value of the rate constant, given that all molecular collisions are successful in contributing to the reaction. The activation energy is the energy barrier that must be overcome for a reaction to occur. Notice that as E_a increases, k decreases. This means that the reaction rate decreases as the energy barrier increases. If we take the natural logarithm of both sides of the Arrhenius equation, we get the equation

$$\ln k = -E_a/RT + \ln A$$

This equation has the form of a straight line. A graph of ln k versus 1/T has a slope of $-E_a/R$ and an intercept of ln A.

Self-Test 3

1. The reaction A → B obeys the Arrhenius equation. The following table shows rate constants for this reaction at various temperatures:

k (s^{-1})	T (K)	1/T (K^{-1})	ln k
2.52×10^{-5}	463		
5.25×10^{-5}	472		
6.30×10^{-4}	504		
3.16×10^{-3}	524		

Fill in the columns 1/T (K^{-1}) and ln k. Now use the following graph to plot ln k versus 1/T, and calculate E_a from your graph:

Workshop: Rate Data and Rate Equations

1. What are the units of the rate constant for a first-order reaction? For a second-order reaction? Explain.

2. Consider the reaction

$$2\ I^-(aq) + S_2O_8^{2-}(aq) \rightarrow I_2(aq) + 2\ SO_4^{2-}(aq)$$

The following initial rates were found from a series of three experiments:

Experiment Number	$[I^-]$ (M)	$[S_2O_8^{2-}]$ (M)	Rate (mol/L · s)
1	0.07500	0.9000	2.61×10^{-4}
2	0.07500	0.4500	1.29×10^{-4}
3	0.1500	0.4500	2.60×10^{-4}

Determine the rate law for the reaction, including the value and units for the rate constant.

3. Instantaneous rates are determined experimentally by the application of a technique known as the *method of initial rates.* The initial instantaneous rate of reaction is approximated by measuring the change in concentration in the smallest time interval that is practical:

$$\text{rate} = \frac{\Delta[A]}{\Delta t} \cong \frac{[A]_{t=1} - [A]_{t=0}}{t_1 - t_0}$$

We can measure the initial rate for different initial concentrations of each of the reactants and, from those data, find the rate order. The value of the rate constant follows because now we know the initial rate, the initial concentrations, and the rate orders for each reactant.

For the reaction A → B, the following concentrations of A were measured:

[A] (M)	time (s)
0.100	0
0.091	1.00
0.082	2.00
0.073	3.00
0.068	4.00
0.061	5.00
0.055	6.00
0.049	7.00
0.045	8.00

a. Use an algebraic approach to determine the initial rate.

b. Assume that the reaction is first order and calculate the rate constant.

c. Assume that the reaction is second order and calculate the rate constant.

d. Divide into pairs. Each pair should choose one of the times listed in the preceding table and calculate the expected concentration on the basis of the initial concentration and rate constant for both first- and second-order reactions. Which rate constant best predicts the actual concentration at each time?

e. Using the data from part (d), plot the following quantities on the graph that follows:

   i. Concentration (y-axis) vs. time (x-axis) for the given experimentally determined data.

   ii. The predicted concentration vs. time for a first-order reaction.

   iii. The predicted concentration vs. time for a second-order reaction.

f. Explain what each of the curves in part (e) tells you about the reaction order.

4. Consider the reaction of the iodide ion and methyl bromide:

$$I^- + CH_3Br \rightarrow CH_3I + Br^-$$

Chemists believe that the only collisions that can result in a reaction are the ones in which the iodide ion approaches the methyl bromide molecule from the opposite side of the location of the bromine atom.

Construct a molecular model of methyl bromide and use it to consider how an effective collision can take place. From your analysis, estimate the fraction of collisions which occur that will have the proper orientation for a successful reaction.

5. The following data were collected for the reaction of iodide ion and methyl bromide (see Question 4):

T (K)	k (L/mol · s)
273	0.0000418
300	0.000860
340	0.00314
370	0.281

a. Determine the activation energy for this reaction. (*Hint:* Question 1 of Self-Test 3 provides one method for determining the activation energy.)

b. The reaction of iodide ion and methyl bromide follows the rate law

$$\text{rate} = k\,[I^-]\,[CH_3Br]$$

If the concentration of each reactant is initially 0.100 M, how long will it take for the concentration of the reactants to drop to 0.050 M if the reaction takes place at 273 K? What if the reaction takes place at 370 K?

c. Determine the percentage increase in reaction rate that occurs as a result of increasing the reaction temperature from 273 K to 370 K.

d. A rule of thumb used by chemists is that the rate of a reaction roughly doubles with each 10-degree increase in temperature. Does the reaction of iodide ion and methyl bromide follow that rule of thumb?

6. Consider again Question 1 of Self-Test 2. Discuss the difference between an instantaneous rate equation and an integrated rate equation. Use graphs or drawings (or both) to illustrate your answer.

The Transition State and Catalysis

"What is now proved, was once only imagined."
William Blake

An old saying among chemists is "A mechanism cannot be proved, only disproved." This may appear to be somewhat of a strange adage. Why did chemists, who felt that they could prove that atoms exist and that carbon forms bonds with tetrahedral angles, feel so inadequate when it came to mechanistic study? The answer probably lies in the fact that more than one mechanism can explain rate data. With only rate data available, it is possible to show that a particular mechanism is inconsistent with the data, but not that it is the only possible mechanism that explains the data.

As experimental technology continues to progress, we develop new instruments that provide detailed structural information on a timescale not imagined just a few years ago. By pursuing a chemical mechanism through multiple methods, a more detailed picture of reaction mechanisms can now be obtained at the submicroscopic level.

The concept of a rate, the distinction between first- and second-order reactions, and the relationship of a rate to equilibrium are fundamental ideas of chemical kinetics. In this workshop, we further explore the connection between the experimental rate observations and what actually occurs in chemical reaction mechanisms at the molecular level.

Self-Test 1

1. Rank each of the statements that follow on a scale from 1 to 10, where 1 is the lowest and 10 is the highest, with regard to the level of proof or evidence you believe they afford. For example, you may rank the statement "I am invisible to other people" as a 1 and the statement "I am alive" as a 10.

a. Michael Jordan was the greatest player in NBA history.
b. The earth is spherical.
c. Smoking cigarettes increases the risk of lung cancer.
d. For every action, there is an equal and opposite reaction.
e. Atoms are composed of electrons, protons, and neutrons.

Reaction Rates

Why are some chemical reactions, such as combustion, extremely fast, while others, such as rusting, so slow? To answer this question, consider the nucleophilic substitution of a chlorine atom in CH_3Br. The reactants have a particular amount of energy. A diagram such as the one in Figure 16.1 can be drawn to indicate the progress of the reaction on the structural level on the x-axis and the energy of a point along the reaction pathway on the y-axis. As the chlorine atom approaches the CH_3Br molecule, the energy increases because of the initial repulsion as the reactants become close. The energy decreases as the bromine atom begins to leave and the incoming chlorine atom begins to bond. The reactant configuration at the peak of the hill is called the *transition-state structure*, and the energy at that point is the *transition-state energy*. The higher the energy of the transition state, the slower is the reaction.

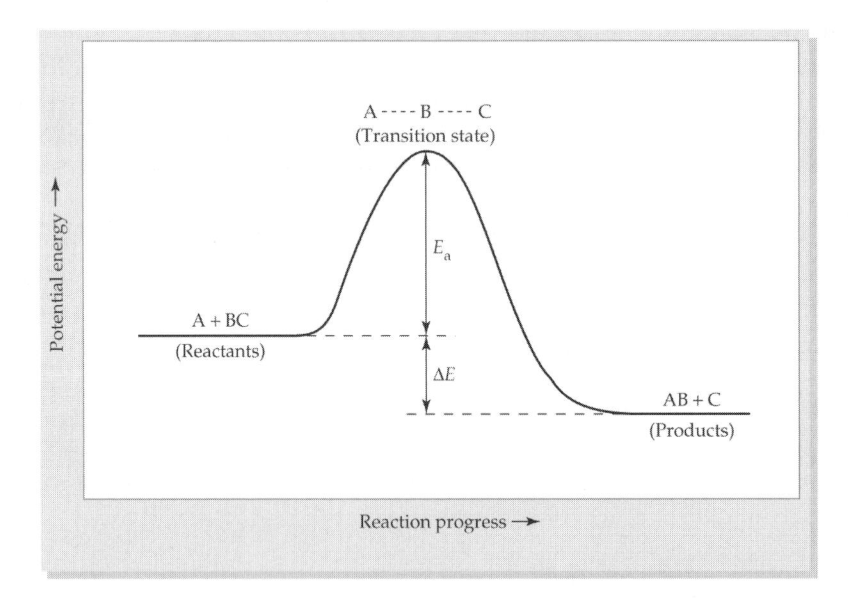

Figure 16.1. A reaction energy diagram for the general exothermic reaction A + BC → AB + C. The reactants, A + BC, increase in potential energy as they approach one another until the maximum energy point, the transition state, is reached. The difference between the energy of the transition state and the initial reactant energy is E_a, the activation energy. The energy decreases as the products, AB + C, are formed from the transition-state structure. The net difference between the energy of reactants and products is given by ΔE.

Reactants can overcome the transition-state energy barrier by having enough energy to equal the transition-state energy, starting from the original reactant configuration. At any given temperature, a collection of molecules will have a distribution of energies and an average energy. As the temperature is raised, more molecules will have sufficient energy to overcome the activation energy barrier.

Transition State Diagrams and Catalysis

If the structure of the transition state is understood, it is known not only how the reaction occurs, but also how to make it proceed faster. Linus Pauling first suggested that the action of enzymes probably occurs through transition-state stabilization. If the energy of the transition state can be lowered by a temporary interaction with another species, the rate of the reaction can be increased. Such a process is called *catalysis.* The initial and final energies of the reactants and products of a catalyzed reaction are exactly the same as those of an uncatalyzed reaction. A catalytic agent does *not* affect the equilibrium position of the reaction; only the reaction pathway is altered.

Reaction Order and Molecularity

The rate order of a reaction is obtained experimentally by varying the concentration of the reactants and observing the effect on the reaction rate. This process does not directly provide any information on the actual molecular-level steps that occur in the reaction, however. The term *molecularity* refers to the number of molecules involved in the rate-determining step of a reaction.

Self-Test 2

1. What is the difference between reaction order and molecularity? Can either be measured? Explain.

Consider a two-step reaction:

Step 1: $A + B \rightleftharpoons C$

Step 2: $C \rightarrow D$

The overall reaction is therefore $A + B \rightarrow D$

We will now examine three conditions under which this two-step reaction can occur and discuss how these conditions are used to find the rate laws.

Case 1: The Steady-State Approximation

This approach is used in many situations in which the concentration of the intermediate C does not change significantly over time. In most reactions with this type of mechanism, the concentration of C initially rises rapidly over a very brief time interval and then remains constant for the duration of the reaction, until one reacting species is completely consumed. When an intermediate species remains at a *steady-state* concentration, we can make the approximation $\Delta[C]/\Delta t = 0$.

The rate at which A and B produce successful collisions determines the speed at which the reaction occurs. This means that the *molecularity* of the reaction is 2, as is the overall rate order.

We begin writing the rate law for the formation of the product D:

$$\text{Rate} = k_2\,[C] \qquad \textbf{(16.1)}$$

This equation says that the rate of the reaction is equal to the rate constant for Step 2, times the concentration of C. However, it is not useful to express a rate law in terms of a short-lived and difficult-to-measure intermediate. Equation 16.1 needs to be expressed in terms of concentrations of the reactants A and B.

The concentration of the intermediate C conforms to the *steady-state approximation:*

$$\frac{\Delta\,[C]}{\Delta t} = 0 \qquad \textbf{(16.2)}$$

Equation 16.2 says that the change in concentration of the intermediate C per unit of time is zero. Given that this is true, the rate of formation of C is equal to the rate of decomposition of C. Let us write the equations for each of these equal rates for C.

The rate of formation of C comes from the forward reaction in Step 1, in which A and B collide to form C:

$$\text{Rate of formation of C} = k_1 [A] [B] \qquad \textbf{(16.3)}$$

C decomposes in two ways: It can undergo the reverse reaction back to A and B in Step 1, and it can react to form D in Step 2. Thus, the rate of decomposition of C will be the sum of the two possible reactions:

$$\text{Rate of decomposition of C} = k_{-1} [C] + k_2 [C] \qquad \textbf{(16.4)}$$

To keep C at steady state, the rate of formation must equal the rate of decomposition:

$$k_1 [A] [B] = k_{-1} [C] + k_2 [C] \qquad \textbf{(16.5)}$$

Equation 16.5 can be solved for the concentration of C, and the result can then be substituted into Equation 16.1 to arrive at a rate-constant expression in terms of the reactants A and B. We leave this for you to complete as a self-test exercise.

Case 2: Preequilibrium

This case is applicable when Step 1 is fast and Step 2 is slow. Step 2 will therefore be the rate-determining step, and the reaction has a molecularity of unity.

The first step in writing the rate law is the same as in Case 1, where we start with the law in terms of the rate of formation of the product, D:

$$\text{Rate} = k_2 [C] \qquad \textbf{(16.1)}$$

Again, we do not want the rate expression in terms of the intermediate C. This time, since both the forward and reverse rates for Step 1 are fast, the reaction in that step will go to equilibrium. We can thus write the equilibrium constant expression for Step 1:

$$K_{eq} = \frac{[C]}{[A] [B]} \qquad \textbf{(16.6)}$$

If Equation 16.6 is solved for the concentration of C and substituted into Equation 16.1, the result is the rate law in terms of the reactants. We leave these final steps for you to complete in the self-test.

Self-Test 3

1. Complete the derivation of the rate laws for Case 1 and Case 2.

2. Draw reaction energy diagrams for Case 1: The Steady-State Approximation and Case 2: Pre-Equilibrium. These are two-step reactions, so there will be two transition-state "bumps" in your diagrams.

Case 3: Pseudo-First Order

A pseudo-first-order reaction is an experimental technique used to simplify the determination of a reaction's kinetics. If we add a large amount of B at the start of the experiment—a technique known as *flooding*—the change in the concentration of B over the course of the reaction will be relatively small.

Our rate law before flooding would depend on the concentrations of both reactants:

$$\text{Rate} = k\,[A]\,[B] \qquad \textbf{(16.7)}$$

However, the concentration of B remains constant because of the pseudo-first-order technique, so it can be absorbed into a new constant

$$k' = k\,[B] \qquad \textbf{(16.8)}$$

Equation 16.7 then becomes

$$\text{Rate} = k'\,[A] \qquad \textbf{(16.9)}$$

In kinetics experiments in general, no matter what type of reaction is being examined, chemists typically begin by measuring and determining rate orders, and then they try to devise mechanisms that are consistent with the experimentally observed rate orders. Unfortunately, neither the molecularity nor the rate order is necessarily given by the overall stoichiometry for a reaction.

Workshop: The Transition State and Catalysis

1. Compare and discuss your rankings from Self-Test 1, Question 1. Begin by reading Statement (a) aloud, and then have each person in your group give his or her rating of its level of proof or evidence. Are there any statements for which the ratings differ by four points or more? What implications do your results have for science in general?

2. a. Draw a transition-state diagram for a one-step reaction in which the forward step is very slow and the reverse step is very fast.

   b. Draw a transition-state diagram for a one-step reaction in which the forward step is very fast and the reverse step is very slow.

   c. Compare the two diagrams from parts a and b. What is the relationship between the relative forward and reverse reaction rates and the initial-reactant and final-product energies?

   d. How are your diagrams related to the relationship $K_{eq} = \frac{k_{forward}}{k_{reverse}}$? In other words, what is the relative magnitude of the equilibrium constants for the two scenarios given in parts a and b?

3. Although many reactions take place in three or more steps, the concepts learned from a study of relatively simple two-step reaction mechanisms are applicable to a wide range of complex reactions. Consider the following two-step mechanism:

Step 1: $A \rightleftharpoons B$

Step 2: $B \rightarrow C$

Divide your group into two subgroups for the next two simulations. Each subgroup should perform both simulations.

Assign one person to keep track of the number of particles of A, another person to count the number of particles of B, a third person to monitor C particles, and a fourth to record the results. The recorder should set up a chart to keep a record of the number of particles of each species after each exchange step. Continue each simulation until you have completed 20 exchange steps.

a. Start with 1000 particles of A. In each exchange step, 10% of the A particles are converted to B. Species B then reacts by giving 10% of its particles back to A and 1% forward to C.

b. Begin with 1000 particles of A. In each step, convert 10% of the A particles to B. Species B then reacts by giving 10% of its particles to A in the reverse reaction and 10% to C in the forward reaction.

c. Compare the results for the simulations between the two subgroups. Reconcile any differences before continuing.

d. Plot the number of particles of B (concentration) versus number of exchange steps (time) for both simulations on the same graph, using the grid that follows. Explain the primary difference.

e. Review the definitions of *steady state* and *preequilibrium* as necessary. How do these concepts apply to the two simulations? What is the relationship between those concepts and relative rates?

f. When the concentration of an intermediate is low and slowly changing, a steady-state approximation can be made. Recall that when this is done, it leads to the fact that the rate of formation of the intermediate is equal to the rate of decomposition. Explain how this concept is related to the simulation in part b.

g. Draw a two-hump transition-state diagram for both simulations.

4. Enzymes typically are proteins that catalyze chemical reactions in living organisms. Interestingly, the names of enzymes often reflect their purpose. For example, the enzyme lactate dehydrogenase literally dehydrogenates (catalyzes an oxidation–reduction reaction of) lactate molecules. Enzymes work by coordinating with a compound in such a way that the transition state is stabilized relative to the reaction that would occur without the presence of the enzyme.

Enzymes are extremely selective, in that a given enzyme catalyzes only one specific chemical reaction. A model used to understand the action of enzymes is called the *lock-and-key mechanism.* The lock is the compound undergoing the reaction, and it is known as the substrate and symbolized S. The key is the enzyme, E. The reaction of substrate to products, $S \rightarrow P$, has a lower transition-state energy when the enzyme is present.

Enzyme kinetics are often characterized by what is known as the *Michaelis–Menten equation,* named after Leonor Michaelis and Maud Menten, who were important early contributors to our understanding of enzyme kinetics. The equation is based on a simple two-step reaction:

$$E + S \rightleftharpoons ES \rightarrow E + P \qquad \textbf{(16.10)}$$

Note that the enzyme is both a reactant and a product of the overall reaction. This property is characteristic of all enzymes.

a. Draw a reaction–energy diagram for an exothermic one-step reaction $S \rightarrow P$. On the same plot, overlay a diagram for the two-step reaction given in Equation 16.10.

b. Derive the rate law for a reaction that follows the Michaelis–Menten mechanism, using the following guidelines:

Step 1: Compare Equation 16.10 to the general two-step reaction given in the introduction to this unit. In both cases, the starting point is Rate $= k_2$ [ES].

Step 2: The total enzyme concentration is equal to the sum of the concentration of the bound enzymes plus the concentration of the unbound enzymes; that is, $[E]_{total} = [ES] + [E]$.

Step 3: Assume that the steady-state approximation is valid for the concentration of ES. Therefore, the rate of decomposition of ES is equal to the rate of formation of ES. Write expressions for each rate, and set them equal to each other. Substitute the equivalent expression from Step 2 for [E].

Step 4: Solve for [ES].

Step 5: Substitute the expression from Step 4 into the expression from Step 1.

Step 6: The Michaelis–Menten equation at this point is usually written as

$$\text{Rate} = \frac{k_2\,[E]_{total}\,[S]}{K_m + [S]}$$

K_m is known as the *Michaelis constant.* What are the constants that make up the Michaelis constant?

c. The most familiar form of the Michaelis–Menten equation substitutes

$$\text{Rate}_{\text{maximum}} = k_2\,[E]_{\text{total}}$$

into the result in Part (b). Make this substitution, and then consider how the reaction rate will be affected when $[S] >> K_m$ and when $[S] << K_m$.

d. Enzyme kinetics are often referred to as *saturation kinetics* because of how the rate changes when the concentration of S is large. What does *saturation* mean in this context? Why does the reaction rate level off when the concentration of S is very large?

e. Derive the enzyme kinetics rate law, using a preequilibrium assumption rather than a steady-state approximation. Historically, this was the original approach to understanding enzyme kinetics.

Equilibrium Concepts

> *"Any change in one of the variables that determines the state of a system in equilibrium causes a shift in the position of the equilibrium in a direction that tends to counteract the change in the variable under consideration."*
> *Henri Le Chatelier*

Some chemical reactions go to completion; that is, the reactants change into products until the reactants are completely consumed. Other reactions are reversible: As products are formed, they react to re-form reactants. For these reversible reactions, an *equilibrium* exists when the rates of the forward and reverse reactions become equal. You probably have already studied two examples of physical equilibrium: the formation of a saturated solution and the equilibrium between a liquid and its vapor. Now we will begin to study the concept of *chemical* equilibrium. Chemical equilibria abound in nature, especially in living organisms. Right now, there are a number of critical chemical equilibrium systems operating in your body that allow you to live. Oxygen molecules bind to hemoglobin molecules in your lungs, where the concentration of oxygen is high, and the oxygen is released in other parts of your body, where oxygen concentration is low. The acidity of your blood is controlled by a chemical equilibrium between carbonic acid and hydrogen carbonate ion.

Self-Test 1

1. When a chemist speaks of an equilibrium, what does he or she mean? Specifically, what is *equal* in an equilibrium?

Liquid–Vapor Equilibrium

Consider a sealed container partially filled with water, as shown in Figure 17.1. We will assume that the container was completely empty before the water was poured in. The rate at which the water evaporates, $H_2O(\ell) \rightarrow H_2O(g)$, depends on the temperature and the surface area of the liquid. If we hold the temperature constant, the rate of evaporation in our container will be constant. This situation is represented by the upward-pointing arrows in the figure. Notice that all upward-pointing arrows have the same size, indicating that the evaporation rate does not change with time.

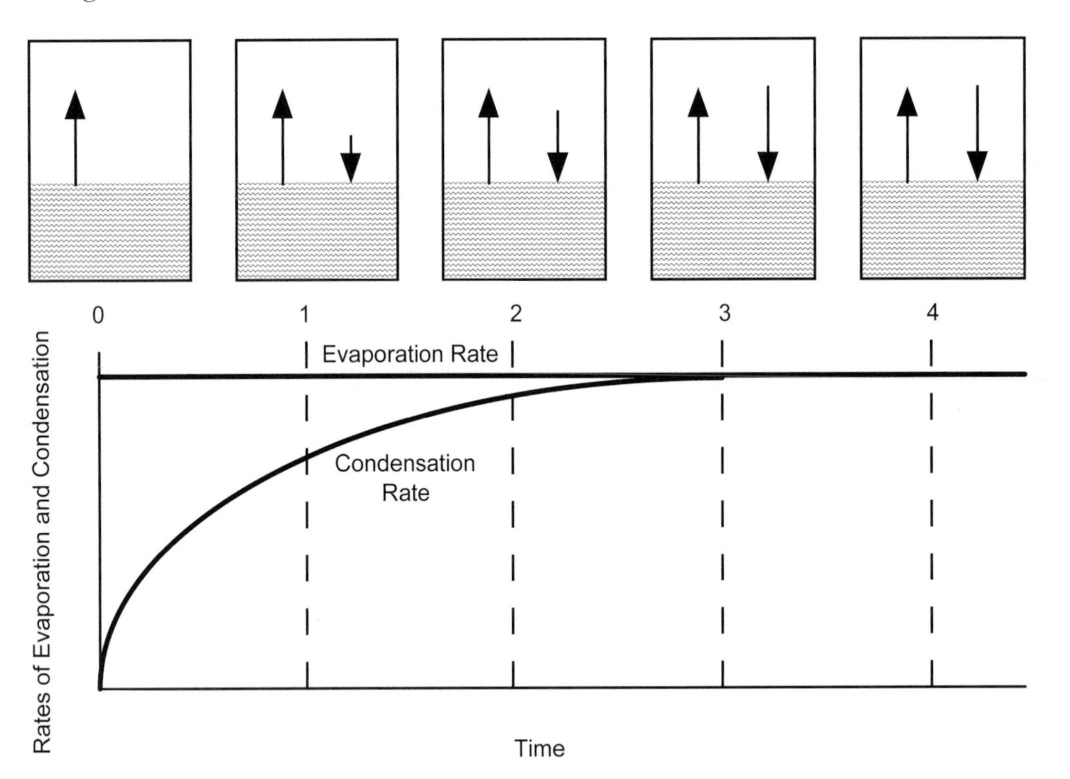

Figure 17.1. Liquid–Vapor Equilibrium. Water is placed in a closed flask and comes to equilibrium with its vapor at constant temperature. Upward-pointing arrows represent the evaporation rate, downward-pointing arrows the condensation rate. Equilibrium is established at Time 3, the time at which the evaporation rate equals the condensation rate.

As the water evaporates, the concentration of water molecules in the vapor phase begins to increase. Some of these gaseous molecules come in contact with the surface of the liquid and reenter the liquid phase, a process known as *condensation*. This process can be represented with the equation $H_2O(g) \rightarrow H_2O(\ell)$. The rate at which it occurs depends on the temperature, the surface area of the liquid, and the concentration of molecules in the vapor phase. Since we are considering a system at constant temperature and liquid surface area, the condensation rate depends only on the vapor concentration. As more water evaporates, the concentration of the vapor phase increases, and the condensation rate increases. This relationship is represented by the downward-pointing arrows in the figure.

When the rate of evaporation is greater than the rate of condensation, the concentration of molecules in the vapor phase will increase, leading in turn to an increase in the condensation rate. The process will continue until the condensation and evaporation rates become equal (Time 3). Particles continue to move between the vapor and liquid phases after the opposing rates become equal, but the number of particles moving from liquid to vapor equals the number moving from vapor to liquid.

When the opposing rates of change become equal, the system is said to be at equilibrium. This kind of equilibrium is called a *dynamic equilibrium,* because, at the submicroscopic level, molecules are in constant motion from one phase to the other. If, however, we consider what is happening at the macroscopic level, we would measure no change: The water vapor pressure remains the same once equilibrium is established. Although this system appears static at the macroscopic level, it is important to keep in mind the dynamic nature of the system at the submicroscopic level.

When a change happens in two directions, such as liquid to vapor and vapor to liquid, the process is said to be *reversible.* If the equilibrium is physical, we refer to a *reversible change.* If we are describing a reversible chemical process, it is a *reversible reaction.* Reversible changes and reversible reactions are represented by equations with double arrows. For the process illustrated, we have $H_2O(\ell) \rightleftharpoons H_2O(g)$.

The Conditions of Equilibrium

The liquid–vapor equilibrium discussed here provides an example of four conditions found in every physical and chemical equilibrium:

1. The opposing rates of change are equal. These are the *only* things that are *equal* in an equilibrium. In particular, note that the amounts of the substances involved in an equilibrium system do *not* have to be equal.

2. The equilibrium is dynamic at the particulate level. Particles are continually changing from what is represented on the left side of the equation to what is represented on the right side, and vice versa. The process will appear static, however, at the macroscopic level.

3. The change must be reversible. An equilibrium cannot be established with a system for which there is no forward *and* reverse reaction or change.

4. The system must be closed. If the container in our liquid–vapor equilibrium was open, many of the gaseous water molecules would leave the vicinity of the container and therefore not be available for the condensation reaction.

Self-Test 2

1. A mountain river flows into a lake. The level of the lake varies with the amount of water flowing in. Water from the lake also evaporates. On a certain day, the rate of flow into the lake from the river was exactly equal to the rate of evaporation. The level of water in the lake remained constant. Did an equilibrium system exist on that day? Explain.

2. Chemical equilibria are referred to as dynamic equilibria. What is meant by the term *dynamic*? Explain your answer on both the submicroscopic and macroscopic levels.

The Equilibrium Constant

Let's consider an experiment designed, in part, to find the value of the equilibrium constant for the reaction $H_2(g) + I_2(g) \rightleftharpoons 2\ HI(g)$.[1] In this experiment, varying amounts of hydrogen and iodine were placed in a closed container at constant temperature and allowed to come to equilibrium. The equilibrium concentrations of all species were then measured. The experiment was repeated many times at a number of temperatures. Experimental runs were also conducted in which hydrogen iodide was placed in the container and allowed to decompose. Data for some of the experimental trials are given in Table 17.1.

Reaction	$[H_2]$	$[I_2]$	$[HI]$	$\frac{[HI]^2}{[H_2]\,[I_2]}$
Combination	0.0024939	0.0025139	0.016952	45.8
Combination	0.0022754	0.0028396	0.017151	45.5
Decomposition	0.0025972	0.0025972	0.017632	46.1
Decomposition	0.0018961	0.0018961	0.012835	45.8

Table 17.1. Equilibrium data for the reaction $H_2(g) + I_2(g) \rightleftharpoons 2\ HI(g)$ at 763.8 K.

An important fact arises from observation of the data in the table: The ratio $\frac{[HI]^2}{[H_2]\,[I_2]}$ has the same value, within limits of experimental variation, for all four trials shown. Many other, similar experiments have been conducted, all reaching the conclusion that the concentration ratio is constant for a particular reaction at a given temperature. This concentration ratio is given the special name *equilibrium constant* and has been assigned the special symbol K. The same conclusion that we have arrived at experimentally can also be derived theoretically. When experimental results are in accord with theory, we have a high degree of confidence that our conclusions are correct.

The equilibrium constant expression is associated with a particular equation. For the general equilibrium $a\,A + b\,B \rightleftharpoons c\,C + d\,D$, the equilibrium constant expression is

$$K = \frac{[C]^c\,[D]^d}{[A]^a\,[B]^b}$$

All equilibrium constant expressions follow this general format. For example, for the decomposition of hydrogen iodide, $2\ HI(g) \rightleftharpoons H_2(g) + I_2(g)$, the equilibrium constant expression is

$$K = \frac{[H_2]\,[I_2]}{[HI]^2}$$

[1]Taylor & Crist, *Journal of the American Chemical Society*, **63**, 1377 (1941).

In the examples we have seen thus far, the reactants have been in the gas phase. Equilibrium constant expressions can be written for reactions in any state of matter, including combinations of states. However, when a liquid solvent is the medium for a reaction, or when a solid is part of a reaction, their concentrations essentially remain constant throughout the reaction. Thus, liquids and solids are not included in equilibrium constant expressions. For example, the reaction $NH_3(aq) + H_2O(\ell) \rightleftharpoons NH_4^+(aq) + OH^-(aq)$ has the equilibrium constant expression

$$K = \frac{[NH_4^+]\,[OH^-]}{[NH_3]}$$

and the reaction $CaCO_3(s) \rightleftharpoons CaO(s) + CO_2(g)$ has the equilibrium constant expression

$$K = [CO_2]$$

You may be wondering about the units on the equilibrium constant. A rigorous mathematical treatment of the term shows that the equilibrium constant is dimensionless. For equilibria at low concentrations, the simpler methods presented here yield the same results as the more complicated techniques.

The Reaction Quotient

The equilibrium constant expression, as its name implies, is valid only for systems at equilibrium. When a system is *not* at equilibrium, we can write and calculate the value of an expression that has the same form as the equilibrium constant. This expression is called *the reaction quotient Q*. By definition, for the general reaction

$$a\,A + b\,B \rightleftharpoons c\,C + d\,D$$

$$Q = \frac{[C]^c\,[D]^d}{[A]^a\,[B]^b}$$

Again, unlike K, Q is *not* restricted to equilibrium conditions.

Self-Test 3

1. Write equilibrium constant expressions for the following reactions:
   a. $PCl_3(g) + Cl_2(g) \rightleftharpoons PCl_5(g)$
   b. $N_2(g) + 3\,H_2(g) \rightleftharpoons 2\,NH_3(g)$
   c. $NH_4HS(s) \rightleftharpoons NH_3(g) + H_2S(g)$
   d. $CaCO_3(s) \rightleftharpoons Ca^{2+}(aq) + CO_3^{2-}(aq)$
   e. $H_2O(\ell) \rightleftharpoons H^+(aq) + OH^-(aq)$

2. What is the difference between K and Q?

Workshop: Equilibrium Concepts

1. The goal of this question is to fill in the blanks in the following statement:

   If a system exists for which $Q < K$, then _____ will happen to bring the system to equilibrium. If a system exists for which $Q > K$, then ____ will happen to bring the system to equilibrium.

   To accomplish this goal, divide into groups of two and consider the data that follow for the equilibrium $N_2O_4(g) \rightleftharpoons 2\ NO_2(g)$, for which $K = 0.00464$ at 25°C. Each pair should analyze one set of data, and then everyone should discuss his or her results with the whole group.

   a. initial moles of $N_2O_4 = 0.250$
   initial moles of $NO_2 = 0$
   container volume = 2.00 L

   b. initial moles of $N_2O_4 = 0$
   initial moles of $NO_2 = 0.500$
   container volume = 1.50 L

   c. initial moles of $N_2O_4 = 0.100$
   initial moles of $NO_2 = 0.100$
   container volume = 0.50 L

   d. initial moles of $N_2O_4 = 0.0500$
   initial moles of $NO_2 = 0.539$
   container volume = 1.00 L

2. *Work in pairs to answer the following questions:*

Consider the hypothetical reversible reaction between two different conformations of a molecule. Conformations are different arrangements of atoms in a molecule that result from rotation around single bonds. An example of such a reversible reaction involves two conformations of cyclohexane and is represented with the following notation:

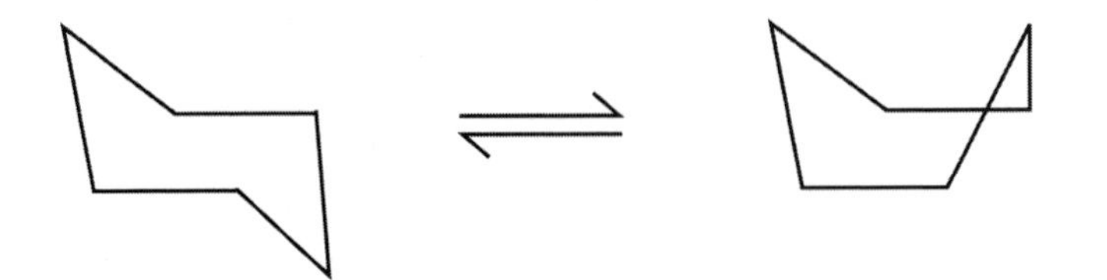

Written in fully, the equation is

$$\text{A (conformation}_1) \rightleftharpoons \text{A (conformation}_2)$$

Each pair should analyze one of the following situations:

a. An equilibrium situation exists with 6 moles of A in conformation$_1$ and 6 moles of A in conformation$_2$.
   i. Two moles of A in conformation$_1$ are removed from the original system. What must happen to restore the system to equilibrium?
   ii. Four moles of A in conformation$_1$ are added to the original system. What must happen to restore the system to equilibrium?
   iii. In each case, how does the initial concentration of reactant molecules compare with the final concentration? How does the initial concentration of product molecules compare with the final concentration? You may find it helpful to make a plot of number of molecules (moles) versus time, specifically illustrating the original equilibrium system, the sudden change that occurs as a result of altering the concentration of one species, and how the system moves back to equilibrium.

b. An equilibrium situation exists with 10 moles of A in conformation_1 and 5 moles of A in conformation_2.

i. Three moles of A in conformation_2 are added to the original system. What must happen to restore the system to equilibrium?

ii. Six moles of A in conformation_1 are removed from the original system. What must happen to restore the system to equilibrium?

iii. In each case, how does the initial concentration of reactant molecules compare with the final concentration? How does the initial concentration of product molecules compare with the final concentration? You may find it helpful to make a plot of number of molecules (moles) versus time, specifically illustrating the original equilibrium system, the sudden change that occurs as a result of altering the concentration of one species, and how the system moves back to equilibrium.

c. An equilibrium situation exists with 6 moles of A in $conformation_1$ and 12 moles of A in $conformation_2$.

i. Three moles of A in $conformation_1$ are removed from the original system. What must happen to restore the system to equilibrium?

ii. Three moles of A in $conformation_2$ are removed from the original system. What must happen to restore the system to equilibrium?

iii. In each case, how does the initial concentration of reactant molecules compare with the final concentration? How does the initial concentration of product molecules compare with the final concentration? You may find it helpful to make a plot of number of molecules (moles) versus time, specifically illustrating the original equilibrium system, the sudden change that occurs as a result of altering the concentration of one species, and how the system moves back to equilibrium.

d. An equilibrium situation exists with 20 moles of A in conformation_1 and 2 moles of A in conformation_2.

i. Eleven moles of A in conformation_1 are removed from the original system. What must happen to restore the system to equilibrium?

ii. Eleven moles of A in conformation_2 are added to the original system. What must happen to restore the system to equilibrium?

iii. In each case, how does the initial concentration of reactant molecules compare with the final concentration? How does the initial concentration of product molecules compare with the final concentration? You may find it helpful to make a plot of number of molecules (moles) versus time, specifically illustrating the original equilibrium system, the sudden change that occurs as a result of altering the concentration of one species, and how the system moves back to equilibrium.

Reassemble as a group, and discuss the effect of a change in concentration on an equilibrium system in terms of how the system responds to reestablish equilibrium. Also, discuss the effects of a change in volume, pressure, and temperature, and the effects of the addition of a catalyst to the system. Write a summary of your results.

Each person in the group should do one step for Questions 3–6.

3. Determine K for the reaction 2 $SO_2(g)$ + $O_2(g)$ $\rightleftharpoons$ 2 $SO_3(g)$, given that 12.12 moles of SO_2 and 7.88 moles of O_2 were placed in a 2.00-liter reaction chamber. The chamber contained 8.05 moles of SO_3 when equilibrium was established.

4. 2.07 moles of oxygen and 2.03 moles of nitrogen monoxide are added to a 0.250-L reaction apparatus. Equilibrium is reached when [NO] = 1.55 M. Calculate K for the equilibrium 2 NO(g) + $O_2(g)$ $\rightleftharpoons$ 2 $NO_2(g)$.

5. $K = 6.9 \times 10^{-3}$ for $N_2O_4(g) \rightleftharpoons 2\ NO_2(g)$ at 300 K. Determine the concentrations of both species at equilibrium if the reaction is initiated by placing 1.00 mole of NO_2 in a 500.0-mL reaction vessel.

6. $K = 8.0$ for the hypothetical reaction $X_2(g) + 3\ Y_2(g) \rightleftharpoons 2\ XY_3(g)$ at a certain temperature. An equilibrium concentration of 0.10 M is needed for $X_2(g)$. How many moles of $XY_3(g)$ must be placed in a 2.00×10^2-mL reaction container to reach the desired concentration of $X_2(g)$?

Introduction to Acids and Bases

> *"For every complex question, there is a simple answer—and it's wrong."*
> *H. L. Mencken*

Originally, the terms *acid* and *base* referred to taste. An acid was something with a sour taste, such as lemon juice, and a base was something with a bitter taste, such as tonic water. It is no coincidence that the acid–base properties of compounds are related to taste. Human taste receptors have evolved to interpret certain molecular features as different tastes. Compounds formed from combinations of acids and bases taste salty and are referred to in chemistry as salts. Sweet compounds have characteristics of both acids and bases in the same molecule.

We will explore the relationship between molecular structure and acid–base properties in detail in Unit 19. In this unit, we will consider only water solutions of acids and bases. A water solution is *acidic* if the hydrogen ion concentration is greater than the hydroxide ion concentration and *basic* if the hydroxide ion concentration is greater than the hydrogen ion concentration. The solution is *neutral* when the hydrogen and hydroxide ion concentrations are equal. In this context, an acid is a substance that increases the concentration of $H^+(aq)$ and a base is a substance that increases the concentration of $OH^-(aq)$.

The Water Equilibrium

Pure water is made up mostly of water molecules, but it also consists of very tiny quantities of hydrogen and hydroxide ions in equal amounts. These ions result from the spontaneous, natural *autoionization* of water:

$$H_2O(\ell) \rightleftharpoons H^+(aq) + OH^-(aq) \qquad \textbf{(18.1a)*}$$

The dissociation of water can also be represented as

$$H_2O(\ell) + H_2O(\ell) \rightleftharpoons H_3O^+(aq) + OH^-(aq) \qquad \textbf{(18.1b)*}$$

*Use Equation 18.1a *or* 18.1b, depending on the preference of your instructor.

We can write an equilibrium constant expression for this reaction in the usual manner. It is assigned the symbol K_w to indicate that it is the equilibrium constant for the autoionization of *w*ater:

$$K_w = [H^+][OH^-] \qquad \textbf{(18.2)}$$

The value of K_w depends on the temperature, as does any equilibrium constant. At 25°C, $K_w = 1.0 \times 10^{-14}$. Since the dissociation of one water molecule yields one hydrogen ion and one hydroxide ion, the concentrations of these ions must be equal in pure water. This relationship allows the calculation of each:

$$[H^+] = [OH^-] = \sqrt{1.0 \times 10^{-14}} = 1.0 \times 10^{-7}\ M \qquad \textbf{(18.3)}$$

Water solutions, composed of solutes dissolved in water, do not necessarily have equal hydrogen ion and hydroxide ion concentrations. Nonetheless, such solutions will still follow Equation 18.2, and the product of the concentrations of the hydrogen and hydroxide ions will equal 1.0×10^{-14} at 25°C.

The pH Scale

Working with the tiny numbers associated with hydrogen ion and hydroxide ion concentrations in solutions can be awkward. For convenience, chemists often work with base-10 logarithms, which are referred to as "p" numbers. In this system, concentrations are expressed as numbers that generally range between 0 and 14. If Z is a value, then, by definition,

$$pZ = -\log Z \qquad \textbf{(18.4)}$$

Applied to hydrogen ion and hydroxide ion concentrations, Equation 18.4 yields

$$pH = -\log [H^+] \quad \text{and} \quad pOH = -\log [OH^-] \qquad \textbf{(18.5)}$$

Working in the other direction, if pH or pOH is known and hydrogen ion or hydroxide concentration is wanted, we have

$$[H^+] = 10^{-pH} \quad \text{and} \quad [OH^-] = 10^{-pOH} \qquad \textbf{(18.6)}$$

Self-Test 1

1. At what pH or range of pH values is a solution considered strongly acidic, mildly acidic, weakly acidic, neutral, weakly basic, mildly basic, and strongly basic? Explain.

2. *Without using a calculator*, determine the pH for each of the following $[H^+]$: 0.010, 1.00×10^{-3}, 1.0×10^{-5}, 1×10^{-8}, 1.000×10^{-9}, 1.0×10^{-11}, 1.0000×10^{-12}. Explain the rules governing pH values and significant figures.

3. Calculate $[H^+]$, $[OH^-]$, and pOH for each of the following pH values: 1.201, 5, 7.25, 10.3, 12.55.

4. An *amphoteric* compound is a compound that is capable of behaving as either an acid or a base. Is water an acid, a base, or amphoteric? Explain. [*Hint:* See Equation 18.1(b).]

Weak Acid Equilibria

A weak acid is a hydrogen-bearing molecular compound that ionizes only slightly in a water solution. Let HA(aq) be the formula for a weak acid dissolved in water. Then the ionization equilibrium is

$$HA(aq) \rightleftharpoons H^+(aq) + A^-(aq) \qquad \textbf{(18.7)}$$

and the associated equilibrium expression is

$$K_a = \frac{[H^+][A^-]}{[HA]} \qquad \textbf{(18.8)}$$

K_a is the *equilibrium constant* for the acid. For typical solutions, the concentration of the undissociated acid, HA, is much greater than the concentrations of hydrogen ion and the anion of the weak acid, A^-. Therefore, weak acid equilibrium constants have values less than unity. The weaker the acid, the smaller is the value of the equilibrium constant. K_a values for many common weak acids are available in reference books and general chemistry textbooks.

Because weak acids ionize only slightly, the amount of weak acid that ionizes is usually negligible compared with the initial concentration of the acid. To illustrate, consider a 0.10-M solution of a weak acid. If this acid is 2.0% ionized at equilibrium, then $0.10 \times 0.020 = 0.0020$ mole of acid per liter of solution is ionized. Applying the rules of significant figures tells us that 0.10 M – 0.0020 M = 0.10 M of acid—which is the initial concentration—remains undissociated. Thus, the amount of acid that ionizes is not significant, and the initial acid concentration is essentially unchanged by the dissociation.

The hydrogen ion concentration and the concentration of the anion of the weak acid cannot be neglected in weak acid equilibria. In fact, determining these concentrations is frequently the goal of both theoretical calculations and experimental investigations of weak acid solutions. If the weak acid solution is not combined with any other compound, then $[H^+] = [A^-]$, because each comes from the same source: the dissociation of HA(aq).

Self-Test 2

1. Write the ionization equation and equilibrium expression for an acetic acid solution. Look up the value of the equilibrium constant and include it as part of your equilibrium expression. Do the same for nitrous acid. Will acetic acid or nitrous acid produce more hydrogen (hydronium) ions in solution? Explain.

Workshop: Introduction to Acids and Bases

1\. a. For each of the following $[H^+]$ values, and without using a calculator, estimate the corresponding pH value: 1.0×10^{-4}, 2.5×10^{-4}, 5.0×10^{-4}, 7.5×10^{-4}, $10. \times 10^{-4}$.

b. Use a calculator to determine the precise pH values corresponding to each estimate in part a. Analyze your estimates, and discuss how you can improve them.

c. For each of the following $[H^+]$ values, and without using a calculator, estimate the corresponding pH value: 1.0×10^{-9}, 2.5×10^{-9}, 5.0×10^{-9}, 7.5×10^{-9}, $10. \times 10^{-9}$.

d. Calculate precise pH values from part c, using a calculator. Analyze your estimates, and discuss how you can improve them.

e. Break into pairs, and then take turns estimating the pH associated with each of the hydrogen ion concentrations that follow. One person in each pair should estimate the pH, and the other person should use a calculator to obtain the precise value. Switch roles back and forth, continuing the exercise until you are confident in your estimating ability.

1.3×10^{-5}, 5.5×10^{-12}, 9.7×10^{-8}, 2.2×10^{-2}, 4.6×10^{-10}, 8.0×10^{-7}.

2. The table that follows includes some of the most common acids and bases. Complete the table, using your textbook as a reference. Recall that strong acids and strong bases are those which are completely dissociated in solution.

Name	Ionization Equation	Strong or Weak/ Acid or Base	K_a or K_b
Hydrochloric acid			
Sodium hydroxide			
Ammonia			
Phosphoric acid			
Nitric acid			
Pyridine			
Carbonic acid			
Water			
Sulfuric acid			
Aniline			

3. Use a penny stacked on a quarter (or any other coin combination or models conveniently available) to represent a monoprotic weak acid molecule. Let the penny be the ionizable hydrogen ion and the quarter be the anion. Ignore water molecules in this exercise.

   a. Starting from 10 un-ionized molecules, develop a model of a solution that is 10% ionized. Is the equilibrium static or dynamic? Explain.

   b. Model a weak acid solution that is 40% ionized.

   c. Model a strong acid solution.

   d. Determine the value of the acid equilibrium constant in each of parts a and b.

4. a. Consider the particle-level composition of a weak acid solution. Ignoring molecular dissociation for now, what is the ratio of weak acid molecules to water molecules in a 0.10-M solution? Determine the answer mathematically, and then draw a proportionally correct sketch of a representative sample of the solution, using open spheres to represent water molecules and filled spheres to represent weak acid molecules. (*Hint:* Water has a density of 1.0 g/mL and a molar mass of 18 g/mol. Your group needs to know the molar concentration of water.)

b. How does molecular dissociation affect the particle-level composition of a weak acid solution? Discuss the answer to this question in terms of your sketch from part a.

c. How does the autoionization of water affect the hydrogen ion concentration of a weak acid solution? (*Hint*: Assume that a typical weak acid has a K_a of about 10^{-5}. What is the K for water?)

d. What is the pH of HCl solutions of concentrations 10^{-3} M, 10^{-6} M, and 10^{-10} M? How does the autoionization of water affect the hydrogen ion concentration of a strong acid solution?

5. Consider an acetic acid solution at equilibrium:

$$HC_2H_3O_2(aq) \rightleftharpoons H^+(aq) + C_2H_3O_2^-(aq)$$

$$K_a = \frac{[H^+][C_2H_3O_2^-]}{[HC_2H_3O_2]}$$

There are three quantities commonly measured or known for a weak acid solution: (i) the value of the equilibrium constant, (ii) the pH of the solution, and (iii) the initial concentration of the acid. In any given situation, one, two, or all three of these quantities may be known, yielding seven possible combinations: i only; ii only, iii only; i and ii; i and iii; ii and iii; and i, ii, and iii.

Consider the table that follows. Check off what is known for each of the seven possibilities. Given each case, what conclusions can be drawn from the known quantities? In other words, what values in the equilibrium constant expression are known for each case?

	K_a	pH	$[HC_2H_3O_2]$	$[H^+]$	$[C_2H_3O_2^-]$
i only					
ii only					
iii only					
i and ii					
i and iii					
ii and iii					
i, ii, and iii					

6. Without making any simplifying assumptions, determine the $[H^+]$ and pH of a 0.0010-M weak acid solution with $K_a = 9.5 \times 10^{-5}$.

7. Assume that, in Question 6, the equilibrium concentration of the acid is the same as the initial concentration. Determine the $[H^+]$ and pH of the weak acid solution. How accurate is your pH when you use this simplifying assumption? How accurate does a pH calculation need to be? Consider a few real-life situations in which pH is an important factor, and discuss the degree of accuracy needed for each situation.

8. Determine the percentage dissociation of each of the weak acid solutions that follow. What is the relationship between concentration and percentage dissociation for a given weak acid?

   a. A 1.0-M solution of a weak acid with $K_a = 5.0 \times 10^{-5}$

   b. A 0.10-M solution of a weak acid with $K_a = 5.0 \times 10^{-5}$

   c. A 0.0010-M solution of a weak acid with $K_a = 5.0 \times 10^{-5}$

9. a. Define the term *hydrolysis*.

b. Write the conventional and net ionic equations for the reaction, if any, that occurs when each of the following are dissolved in water: sodium hydroxide, sodium acetate, ammonia. Are any of these hydrolysis reactions? List the major and minor species in solution for each.

c. Draw a series of illustrations that show what occurs at the particulate level (i) before solid sodium acetate and water are combined, (ii) just after the sodium acetate is added to the water, and (iii) after the solution comes to equilibrium.

10. Compare the similarities and differences in the *strategy* you use to find the pH of each of the solutions that follow. Write a brief summary of the procedure used in each case.

    a. 0.10 M sodium hydroxide

    b. 0.10 M sodium acetate

    c. 0.10 M ammonia

    d. 0.10 M hydrochloric acid

The Acid–Base Concept

> *". . . The concepts of acids and bases are in fact of such a general character that we must consider it a necessary requirement of these concepts in general to formulate a pattern independent of the nature of an arbitrary solvent."*
> *J. N. Brønsted*

In this unit, we take another look at the definitions of acids and bases. We'll examine acid–base behavior in terms of molecular structure, bonding, and chemical reactions with other substances. As we've seen in previous units, acids and bases can be recognized and described by their properties, such as taste, the ability to cause litmus paper to change color, and pH. In this unit, you will learn how these properties relate to composition, molecular structure, and chemical reactions with other substances.

Self-Test 1

1. Complete the table that follows for the acids and bases found in some common substances. Assume that the solutions are 0.1 M. *Use your textbook as a reference.*

Species	Taste	Red Litmus	Blue Litmus	Solution pH	Description
HCl	sour	red	red	1.0	strong acid
NaOH				13.0	
H_2O				7.0	
N_2H_4				10.0	
Na_2CO_3				11.0	
H_2SO_4				2.0	
CH_3COOH				3.4	

Three Definitions of Acids and Bases

Arrhenius The Arrhenius definition of acids and bases derives from his theories concerning the formation of ions in aqueous solution. Arrhenius first proposed this idea in 1884 as a student, but his professors considered it to be nonsense. As a result, he nearly failed to earn the doctorate. He stuck to his convictions, however, and earned a Nobel Prize in 1903 for his electrolytic theory of dissociation.

In an extension of his ionic theory, Arrhenius proposed that acids were substances that yield hydrogen ions in aqueous solution while bases were substances that produced hydroxide ions in aqueous solution. Arrhenius also believed that an acid must have at least one hydrogen atom and a base must have at least one hydroxyl group. Moreover, the Arrhenius definitions apply only to the behavior of substances in water, as in the following reactions:

Acid: $HCl(aq) \rightarrow H^+(aq) + Cl^-(aq)$

Base: $NaOH(aq) \rightarrow Na^+(aq) + OH^-(aq)$

Brønsted–Lowry The Brønsted–Lowry definition of acids and bases liberates the acid–base concept from its limitation to aqueous solutions, as well as from the requirement that bases contain the hydroxyl group. A *Brønsted–Lowry acid* is a hydrogen-containing species that is capable of acting as a proton (hydrogen ion) donor, while a *Brønsted–Lowry base* is a species that is capable of acting as a proton acceptor. The following reaction is illustrative:

$$HCl(g) + H_2O(\ell) \rightarrow H_3O^+(aq) + Cl^-(aq)$$

We see that the Brønsted–Lowry acids and bases include those substances which are classified as Arrhenius acids and bases. A substance that yields hydrogen ions is a proton donor, and a substance containing a hydroxyl group that is capable of yielding hydroxide ions in aqueous solution would be a proton acceptor. Brønsted–Lowry bases, however, are required neither to contain hydroxyl groups nor to form hydroxyl ions in solution, as in the following reaction:

$$Cl\text{–}H + :NH_3 \rightarrow [NH_4]^+ [Cl]^-$$

Cl–H	:NH₃
proton donor	*proton acceptor*

Lewis The Lewis definition is the most general of the three. It liberates the acid–base concept from its reliance on the presence of any particular element, focusing instead on the behavior of the electrons during an acid–base reaction. The importance of the Lewis definition is that it gets at the basis of acid–base behavior and catalogues the largest number of molecules and reactions. A *Lewis acid* is an electron pair acceptor; a *Lewis base* is an electron pair donor. The following reaction illustrates the two concepts:

$$H_3N: + BF_3 \rightarrow H_3N:BF_3$$

$H_3N:$	+	BF_3	→	$H_3N:BF_3$
acid		*base*		
electron pair donor		electron pair acceptor		

Self-Test 2

1. For each of the reactions that follow, classify the *first reactant listed*—the one in boldface type—as an Arrhenius acid, an Arrhenius base, a Brønsted–Lowry acid, a Brønsted–Lowry base, a Lewis acid, or a Lewis base. More than one classification may apply.

   a. $\mathbf{SO_3} + O^{2-} \rightarrow SO_4^{2-}$

   b. $\mathbf{CN^-} + H_2O \rightarrow HCN + OH^-$

   c. $\mathbf{B(OH)_3} + 2\,H_2O \rightarrow B(OH)_4^- + H_3O^+$

   d. $2\,\mathbf{H_2O} \rightarrow H_3O^+ + OH^-$

   e. $\mathbf{H_2O} + HCl \rightarrow H_3O^+ + Cl^-$

   f. $4\,\mathbf{LiH} + AlCl_3 \rightarrow LiAlH_4 + 3\,LiCl$

Conjugate Acids and Bases

The concept of conjugate acids and bases is best understood by considering what happens when a substance behaves as a Brønsted–Lowry acid or a Brønsted–Lowry base. By reversing the reaction in which a substance acts as a proton donor, we see that the product is itself a proton acceptor. It is thus a base—or, more specifically, it is the conjugate base of the original acid. Similarly, when a substance behaves as a Brønsted–Lowry base, the product is a proton donor. It is thus the conjugate acid of the original base. In sum, a *conjugate base* is the species that remains after a Brønsted–Lowry acid donates a proton, and a *conjugate acid* is the species that forms when a Brønsted–Lowry base accepts a proton:

$H_2O(\ell)$	+	$NH_3(aq)$	$\rightarrow$	$NH_4^+(aq)$	+	$OH^-(aq)$
acid	+	*base*	$\rightarrow$	*conjugate acid*	+	*conjugate base*
donates proton		*accepts proton*				

Properties of Atoms and Molecules

Whether a substance is classified as an acid or a base, the strength of that acid or base can be determined by examining a sample of the substance in some manner. For example, we could use taste to measure acidity, with an acid so classified if it is sour, and the degree of sourness to measure the strength of acids. As scientists, chemists find it particularly important to be able to obtain such information from the structure of the molecule and the atoms from which it is formed. It is useful, therefore, to review several properties of atoms and molecules that will, in turn, help you to understand what makes a molecule an acid or a base.

Atomic charge is the charge, usually fractional, that an atom or a group of atoms carries when it is in a molecule. All of the atomic charges in a species must add up to the charge on that species. Atomic charges are important because the charge on an atom in a molecule is its optimal charge. In a reaction, if this charge is to be increased or decreased, energy must be supplied.

Electronegativity is the ability of an atom to attract electrons to itself in a bond. Thus, atoms of high electronegativity will acquire excess negative charge (electrons) and will therefore have negative atomic charges. In contrast, atoms of lower electronegativity will have lost negative charge—electrons—and will have positive atomic charges.

A **lone pair** is two electrons that are paired in the same orbital and do not participate in forming a bond.

A **functional group** is an atom or a group of atoms, occurring in an organic molecule, that bestows certain chemical properties on that molecule. For example, the –COOH functional group occurs in organic acids and is the source of acidity, regardless of the nature of the hydrocarbon backbone of the molecule.

pH is a logarithmic unit of concentration. Specifically, $pH = -\log [H^+]$.

Self-Test 3

1. Rank the following in order of increasing electronegativity:
Na, Cl, K, Ba, F, O, C, B, N, Br, Be

2. Rank the following in order of increasingly positive hydrogen charge:
   a. HF, LiH, CH_4, H_2O, CsH

   b. OClOH, ClOH, IOH, O_3ClOH, O_2ClOH

3. Rank the following in order of increasingly negative hydroxyl group charge:
NaOH, CsOH, CH_3OH, ClOH, LiOH

Workshop: The Acid–Base Concept

Instructions: Divide your workshop group into two subgroups. The goal for each subgroup is to develop as many rules or guidelines for predicting the acid–base character (including strength) of substances *from a knowledge of their molecular structure and the properties of their atoms*. Once you have generated a list of rules, rank them in order of least tentative to most tentative and then in order of most general to least general. Use the observations that follow as the basis for developing your rules. Be sure that the rules you develop hold not only for the data used in developing them, but also for other experiments and different examples. At the end of the allotted time, each subgroup will present its rules and defend them against criticism by the other subgroup.

Subgroup Organization: Assign the roles that follow to members of your subgroup. You will not have enough people in your subgroup to fill all of the roles, so some people will have to double up.

Manager: Manages the process from start to finish and acts as the group strategist.

Timekeeper: Budgets time for each task and ensures completion of the workshop by the deadline.

Spokesperson: Speaks on behalf of the whole group.

Recorder: Keeps records of the group's deliberations and conclusions.

Skeptic: Plays the role of "devil's advocate" and critiques the conclusions of the group.

Example: Use this example as a guide to carry out the instructions for completing this workshop. One series of interactions among the members of your group may adhere to the following format:

Spokesperson: The tentative rule is "Compounds that contain halogen atoms bonded to oxygen are strong acids."

Skeptic: Is this rule general for all halogens? What about hydrogen fluoride, which is a weak acid?

Group Response: Hydrogen fluoride is an exception. Fluorine is the most electronegative element, and it holds its hydrogen atom so tightly that the molecule doesn't ionize much in water.

Observation Set 1

It was not uncommon for early chemists to characterize substances on the basis of how they tasted. This approach is, of course, extremely dangerous. The table that follows lists the results that one might find if he or she were to dissolve each substance in water. (*Hint*: Draw Lewis structures for each.)

H-Containing Molecules

Formula	Taste
HCl	Sour
CH_4	Neither sour nor bitter
NaH	Bitter
MgH_2	Bitter
AlH_3	Bitter
GeH_4	Neither sour nor bitter
BeH_2	Bitter
HF	Sour
BaH_2	Bitter
SrH_2	Bitter
H_2S	Sour
LiH	Bitter

OH-Containing Molecules

Formula	Taste
$HClO$	Sour
H_2O	Neither sour nor bitter
CH_3CH_2OH	Neither sour nor bitter
$NaOH$	Bitter
$HBrO$	Sour
$HClO_4$	Sour
$LiOH$	Bitter
H_3PO_4	Sour
HNO_2	Sour
$Mg(OH)_2$	Bitter
H_2SO_3	Sour
$Al(OH)_3$	Bitter

Observation Set 2

The characterization of substances by taste was unsatisfactory for a number of obvious reasons. A better method involves the use of acid–base indicators, substances that change color on the basis of the acidity or basicity of the solution. The earliest known acid–base indicators in the history of chemistry were plant extracts, such as syrup of violets or lilacs. These indicators were later absorbed onto paper and dried to yield indicator papers. Among the most common of modern indicator papers are red and blue litmus paper. Red litmus turns blue in basic solution and blue litmus turns red in acidic solution. The table that follows lists results from dissolving each substance in water and performing a litmus test on the solution. Again, Lewis structures will be helpful.

H-Containing Molecules

Formula	Red Litmus	Blue Litmus
HCl	Red	Red
CH_4	Red	Blue
NaH	Blue	Blue
HI	Red	Red
H_2O	Red	Blue
SrH_2	Blue	Blue
H_2S	Red	Red

OH-Containing Molecules

Formula	Red Litmus	Blue Litmus
$HClO$	Red	Red
$Al(OH)_3$	Blue	Blue
CH_3OH	Red	Blue
H_3PO_4	Red	Red
CH_3CH_2OH	Red	Blue
$LiOH$	Blue	Blue
HNO_2	Red	Red

Observation Set 2 (Continued)

H-Containing Molecules | **OH-Containing Molecules**

Formula	Red Litmus	Blue Litmus	Formula	Red Litmus	Blue Litmus
KH	Blue	Blue	HCOOH	Red	Red
HF	Red	Red	$Be(OH)_2$	Blue	Blue
HBr	Red	Red	H_2SO_3	Red	Red
CH_2O	Red	Blue	$OCHCH_2OH$	Red	Blue

Observation Set 3

Improving upon the capabilities of indicator paper, pH meters are able to quantitatively measure the acidity of a solution. All of the pH values reported in the table that follows are measurements of 0.010-M solutions. For each ion, the value represents a solution of the sodium salt of the ion. Once again, drawing Lewis structures will help you see the structure–pH relationship more clearly.

Species	pH	Species	pH	Species	pH
$HClO_4$	2.00	SO_3^{2-}	9.60	H_2S	4.50
H_2SO_4	2.00	HCl	2.00	HPO_4^{2-}	9.60
H_3PO_4	2.24	$H_2PO_4^-$	4.60	HBr	2.00
ClO^-	9.73	HI	2.00	SO_4^{2-}	7.00
ClO_4^-	7.00	H_2SO_3	2.18	PO_4^{3-}	11.89
HSO_4^-	2.88	CH_3COOH	3.37	NaOH	12.00
HClO	4.73	HS^-	9.50	HNO_2	2.72
HSO_3^-	4.60	H_2O	7.00	Cl^-	7.00

Observation Set 4

Atomic charge and electronegativity play an important role in determining the acid–base properties of molecules. These hydrogen and hydroxyl charges have been determined on the basis of the electronegativities of the atoms.

Species	H Charge	OH Charge	Species	H Charge	OH Charge
CH_4	+0.012	na*	NaH	–0.25	na*
HClO	+0.54	+0.22	H_2O	+0.26	–0.26
HSO_4^-	+0.465	+0.099	H_3O^+	+0.43	+0.15
H_2SO_4	+0.53	+0.21	$HClO_4$	+0.67	+0.44
CH_3COOH	+0.35 (O) +0.099 (C)	–0.085	HSO_4^-	+0.41	+0.004
H_3PO_4	+0.36	–0.083	H_2S	+0.20	na*

* not applicable

Observation Set 4 (Continued)

Species	H Charge	OH Charge	Species	H Charge	OH Charge
$Al(OH)_3$	+0.23	–0.23	AlH_3	–0.23	na*
CH_3OH	+0.30 (O) +0.050 (H)	–0.20	LiH	–0.24	na*
$H_2PO_4^-$	+0.36	–0.083	HPO_4^-	+0.26	+0.26
CH_2O	+0.12	na*	NH_3	+0.11	na*

* not applicable

Buffers and Titrations

> *"Biological control of the pH of cells and body fluids is of central importance in all aspects of intermediary metabolism and cellular function."*
> *Albert L. Lehninger*

A *buffer* is a solution that resists changes in pH when a small amount of acid or base is added to it. A buffer system forms when a weak acid and its conjugate base are present in the same solution. Biological solutions, such as blood, are usually buffers. The control of pH is essential to the proper functioning of these solutions. For example, blood must maintain a pH of 7.4 in order to carry oxygen to the cells. A change in pH of 10 percent is enough to destroy the capacity of blood to transport oxygen. As a buffer, blood contains the carbonic acid buffer system, H_2CO_3 and HCO_3^-, in addition to other conjugate acid–base pairs.

Buffer systems are an important application of acid–base equilibria, the study of which is useful because many other chemical systems can be understood through the same mathematical approach. The most common experimental method used to study acid–base systems is titration analysis, through which we can determine the pK_a of a weak acid and the pK_b of its conjugate base, the two essential components of a buffer ($pK_a = -\log K_a$ and $pK_b = -\log K_b$).

The Buffer Equation

Let's consider a weak-acid equilibrium system and its corresponding equilibrium constant:

$$HA(aq) \rightleftharpoons H^+(aq) + A^-(aq) \qquad K_a = \frac{[H^+][A^-]}{[HA]}$$

HA represents a weak monoprotic acid, and A^- is its conjugate base. Solving the equilibrium constant expression for the hydrogen ion concentration yields

$$[H^+] = K_a \frac{[HA]}{[A^-]}$$

Taking the logarithm of each side and multiplying by –1 results in

$$-\log [H^+] = -\log \left(K_a \frac{[HA]}{[A^-]} \right)$$

Algebraically rearranging terms gives

$$-\log [H^+] = -\log K_a - \log \frac{[HA]}{[A^-]}$$

$$-\log [H^+] = -\log K_a + \log \frac{[A^-]}{[HA]}$$

Using the fact that $pQ = -\log Q$, we arrive at the Henderson–Hasselbach equation, which is used to calculate the pH of buffers:

$$pH = pK_a + \log \frac{[A^-]}{[HA]}$$

Since this equation is derived from the equilibrium constant expression, all concentrations must be equilibrium concentrations. In using the equation, it is often useful and accurate to make the approximation that the weak acid is only slightly dissociated. In this case, the equilibrium concentration of HA is approximately equal to the initial concentration, or

$$[HA]_{equilibrium} = [HA]_{initial}$$

A similar assumption is also valid for weak bases:

$$[A^-]_{equilibrium} = [A^-]_{initial}$$

The Effectiveness of a Buffer

In a 100.0-mL solution containing 0.010 mol acetic acid ($HC_2H_3O_2$) and 0.010 mol sodium acetate ($NaC_2H_3O_2$) $[A^-] = [HA]$ and $\frac{[A^-]}{[HA]} = 1$. Since $\log 1 = 0$ and pK_a for acetic acid is 4.75, it follows that $pH = pK_a = 4.75$ for this buffer.

Consider what happens if we add 0.005 mol of HCl to this solution. 0.005 mole of H^+ will react with 0.005 mole of acetate ion, and the number of moles of acetic acid and acetate ion change as follows:

	$H^+(aq)$ +	$C_2H_3O_2^-(aq)$ ⇌	$HC_2H_3O_2(aq)$
Initial Moles	0.005	0.010	0.010
Change	– 0.005	– 0.005	+ 0.005
Final Moles	0	0.005	0.015

The buffer equation can now be applied to determine the new pH of the solution:

$$pH = pK_a + \log\frac{[A^-]}{[HA]} = 4.75 + \log\frac{(0.005\ \text{mol}/0.1000\ \text{L})}{(0.015\ \text{mol}/0.1000\ \text{L})} = 4.27$$

The pH of the solution changes from 4.75 to 4.27 upon addition of the acid. Let's compare this with what will happen if we add the same amount of HCl to a nonbuffered solution that begins at pH = 4.75. A 1.8×10^{-5}-M HCl solution has a pH of 4.75. The number of moles of $H^+(aq)$ in this solution is

$$0.1000\ \text{L} \times \frac{1.8 \times 10^{-5}\ \text{mol}}{\text{L}} = 1.8 \times 10^{-6}\ \text{mol}$$

The amount of HCl added was 0.005 mol, so after the acid is added, the number of moles is 0.005 + 0.0000018 = 0.005 mol. The new hydrogen ion concentration is therefore

$$\frac{0.005\ \text{mol}}{0.1000\ \text{L}} = 0.05\ \text{M}$$

and the pH of the solution is

$$pH = -\log[H^+] = -\log(0.05) = 1.3$$

In this unbuffered solution, the pH changes from 4.75 to 1.3, which is a much larger change than in the buffered solution.

A consideration that must be made in preparing a buffer is to have sufficient quantities of both the weak acid and its conjugate base react completely with any base or acid that may be added to the system. The buffer capacity of a system is defined in terms of the concentrations of the acid–base conjugate pair. Greater concentrations will withstand greater additions of base or acid while still resisting a significant change in pH. If we were to add so much acid that it reacted with all of the base in a buffer system, the buffering capacity of the system would be exceeded, and further additions of acid would result in large changes in pH.

A guideline for preparing a buffer system is to choose an acid with a pK_a within one pH unit of the desired buffer. Doing this ensures that the ratio of base to acid will range between 1 to 10 and 10 to 1; thus, sufficient quantities of both acid and base will be present in the buffering system.

Self-Test 1

1. Consider the following weak acids and their K_a values:

Acetic acid	$K_a = 1.8 \times 10^{-5}$
Phosphoric acid	$K_{a1} = 7.5 \times 10^{-3}$
Hypochlorous acid	$K_a = 3.5 \times 10^{-8}$

You are to prepare buffers at pH = 2.8, 4.5, and 7.5. What weak acid–conjugate base buffer system is the best choice for each pH from the acids listed? Explain your reasoning.

2. Consider again the 100.0-mL buffer containing 0.010 mol acetic acid ($HC_2H_3O_2$) and 0.010 mol sodium acetate ($NaC_2H_3O_2$) that was introduced in the previous section. Determine the resulting pH if 0.005 mol NaOH is added to the buffer.

pH Titration

A pH titration is performed by adding small amounts of a titrant to a solution and simultaneously monitoring the pH of the solution. In a typical titration, small amounts of sodium hydroxide solution are added to a weak acid solution. In this case, the pH during the titration is related to the pK_a of the weak acid.

During the titration, acid is converted to its conjugate base and a buffer solution is formed. Eventually, the quantity of base added is such that all of the acid is converted to its conjugate base and the equivalence point of the titration has been reached. At the equivalence point, the solution is no longer a buffer.

You will investigate the uses and limitations of the buffer equation in this workshop, and you will apply what you already know about changes in pH to evaluate the titration process.

Self-Test 2

1. Consider the titration of an acetic acid solution with sodium hydroxide solution. Specifically, describe the composition of the resulting solution at the following three stages of the titration: (i) before the titration begins, (ii) when the number of moles of sodium hydroxide added is equal to half of the number of moles of acetic acid originally in the beaker, and (iii) when the number of moles of NaOH added is equal to the number of moles of acetic acid originally in the beaker (the equivalence point). Answer and explain your reasoning for each of the following questions:

    a. When does the solution contain mostly acetate ion? When does it contain mostly acetic acid? When does it contain significant amounts of both?

    b. At what point during the titration is the pH of the solution at its lowest value? At what point is it at its highest value? When is it between the two extreme values?

Workshop: Buffers and Titrations

1. Student A claims that she can calculate the pH of a buffer system without knowing the actual concentrations of the acid and conjugate base. Student B disagrees, citing the fact that the buffer equation clearly requires concentrations. Who is correct? Explain.

2. The carbonate buffer system is important in regulating blood pH levels. Carbonic acid is diprotic and therefore has two K_a values:

$$H_2CO_3(aq) \rightleftharpoons H^+(aq) + HCO_3^-(aq) \qquad K_{a1} = 4.3 \times 10^{-7}$$
$$HCO_3^-(aq) \rightleftharpoons H^+(aq) + CO_3^{2-}(aq) \qquad K_{a2} = 5.6 \times 10^{-11}$$

Since the second dissociation has a K_a value significantly smaller than that of the first dissociation, the second dissociation can be assumed to have no effect on the $H_2CO_3(aq)/HCO_3^-(aq)$ equilibrium.

The pH of blood is 7.4. What is the ratio of carbonic acid to bicarbonate ion in blood?

3. Biochemical experiments frequently utilize a buffer system based on *tris*-(hydroxymethyl)aminomethane ($(HOCH_2)_3CNH_2$), which is also called TRIS or THAM. The pK_a of the conjugate acid of TRIS, $(HOCH_2)_3CNH_3^+$, is 8.075. What mole ratio of acid to base is required to prepare a buffer at the same pH as human blood (pH = 7.4)?

4. When a strong base is gradually added dropwise to a weak acid, the pH changes at each addition. When enough base has been added to react with all of the acid, the pH changes sharply, indicating the end point of the titration. A plot of pH versus volume of base added gives what is known as a titration curve.

Consider the titration of 25.00 mL of 0.1000 M acetic acid with 0.1000 M NaOH.

a. Write the equation for the titration reaction.

b. Determine the volume of NaOH solution required to reach the end point.

c. The table that follows has entries for several steps along the titration curve. To calculate the pH at each step, you first must understand the chemistry at that step. Then you can decide the appropriate method to calculate the pH. For each volume listed, (i) list the major species in solution, (ii) determine whether the K_a equation, K_b equation, buffer equation, or solution equilibrium equation is appropriate for the calculation of the solution $[H^+]$ and pH, and (iii) complete the calculations.

Volume NaOH Added (mL)	Major Species	Appropriate Equation	$[H^+]$	pH
0.00				
5.00				
12.50				
20.00				
25.00				
30.00				

5. It is common for a chemist to prepare a buffer by adding sodium hydroxide solution to a weak acid solution, using a pH meter to monitor the pH until the desired buffer pH is reached. This technique accounts for nonideal conditions that can result in small, but significant, errors in theoretically based calculations. How many drops of 1.0 M NaOH should be added to 200.0 mL of 0.050 M acetic acid with $K_a = 1.8 \times 10^{-5}$ to make a buffer with pH = 5.0? Assume that each drop delivers an average volume of 0.05 mL.

6. A scientist extracts and purifies a liquid compound from a plant. An initial test indicates that the compound is a monoprotic weak acid. The scientist dilutes 5.00 mL of the liquid to 50.00 mL and titrates the solution with 0.010 M NaOH. The titration curve is as follows:

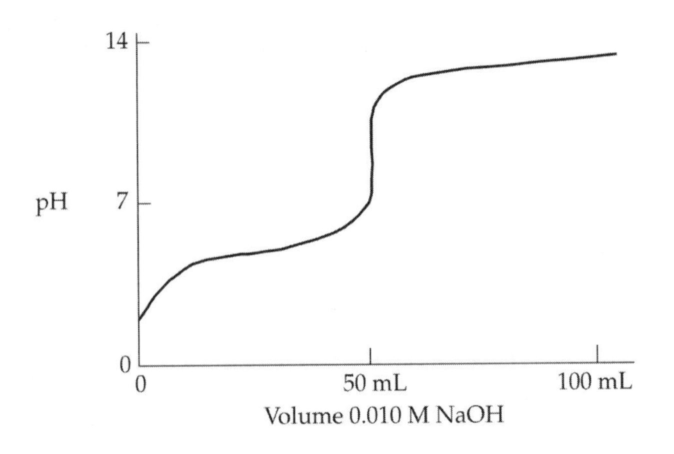

a. How many moles of the compound were in the sample?

b. What is the pK_a of the compound?

Solubility Equilibria

> *"Every electrolyte (in aqueous solution), consists partly of active (in electrical and chemical relation), and partly of inactive molecules . . ."*
> *Svante Arrhenius*

Pure water contains 1.0×10^{-7} mol/liter of hydrogen and hydroxide ions in addition to water molecules. Although that tiny amount may seem insignificant, variations in ion concentrations can result in major changes in many important properties of a solution. For example, the hydrogen ion concentration in blood is normally 3.5×10^{-8} mol/liter, but if it drops to 1.0×10^{-8} mol/liter, death results in only a few seconds. Even though these numbers are small and the change seems small (2.5×10^{-8} mol/liter), this is a very large percentage change (71%). Clearly, an understanding of such tiny concentrations is an important aspect of chemistry.

A similar situation exists in solutions of so-called insoluble salts. Tiny amounts of the insoluble salt dissolve in water, and the solution contains tiny amounts of the ions that make up the salt. For example, even though silver chloride is classified as insoluble, a tiny concentration (9×10^{-5} mol/liter) of silver and chloride ions exists in a solution of the compound. These small concentrations of dissolved ions can be important to industrial and analytical chemists, as well as to physicians and geologists.

The Solubility Product Constant

When silver chloride is added to water, some of the compound will dissolve. When the process reaches equilibrium, the macroscopic concentrations of silver and chloride ions will remain constant, and the process can be represented by the equation

$$AgCl(s) \rightleftharpoons Ag^{+}(aq) + Cl^{-}(aq)$$

Recalling that solids are not included in an equilibrium constant expression, we can write the K expression for this reaction as

$$K_{sp} = [Ag^{+}]\,[Cl^{-}]$$

In this expression, the subscript "sp" on the K refers to the term "solubility product." K_{sp} is called the *solubility product constant.*

K_{sp} expressions are written in the same manner as is any other equilibrium expression. Just keep in mind that solids are not included in those expressions. If we have a solution with lead(II) chloride, the dissolving equation and solubility product constant expression are, respectively,

$$PbCl_2(s) \rightleftharpoons Pb^{2+}(aq) + 2\ Cl^-(aq)$$

and

$$K_{sp} = [Pb^{2+}]\ [Cl^-]^2$$

Recall that ion concentrations are raised to the power of their stoichiometric coefficient in equilibrium constant expressions.

As with all equilibrium constant expressions, the solubility product constant is temperature dependent. We will restrict our interest to reactions occurring at 25°C, the standard thermodynamic temperature, in this unit.

Determination of Solubility Product Constant Values

If we know the solubility of a salt in water, we can calculate its K_{sp} value. We follow a four-step procedure:

1. Write the balanced equation representing the process of dissolution.
2. Write the K_{sp} expression, based on the equation from Step 1.
3. Calculate the mole-per-liter (M) concentration of each of the ions represented in the K_{sp} expression.
4. Substitute the ion concentrations into the K_{sp} expression and calculate K_{sp}.

EXAMPLE 21.1

It is experimentally determined that 1.3×10^{-5} g of silver bromide will dissolve in 100.0 mL of water at 25°C. What is the solubility product constant for silver bromide?

SOLUTION

The first step is to write the balanced equation for dissolving silver bromide:

$$AgBr(s) \rightleftharpoons Ag^+(aq) + Br^-(aq)$$

Now we write the solubility product constant expression:

$$K_{sp} = [Ag^+]\ [Br^-]$$

The third step requires changing the concentration in g/100 mL to mol/L:

$$\frac{1.3 \times 10^{-5}\text{ g AgBr}}{100\text{ mL}} \times \frac{1\text{ mol AgBr}}{187.8\text{ g AgBr}} \times \frac{1\text{ mol Ag}^+}{1\text{ mol AgBr}} \times \frac{1000\text{ mL}}{\text{L}} =$$

$$\frac{6.9 \times 10^{-7}\text{ mol Ag}^+}{\text{L}} = 6.9 \times 10^{-7}\text{ M Ag}^+$$

Since one bromide ion is formed for each silver ion formed, $[Br^-] = [Ag^+] = 6.9 \times 10^{-7}$ M.

Now we can substitute the molar concentrations into the expression for K_{sp} and solve:

$$K_{sp} = [Ag^+]\,[Br^-] = (6.9 \times 10^{-7})\,(6.9 \times 10^{-7}) = 4.8 \times 10^{-13}$$

Self-Test 1

1. An experiment is conducted to determine the solubility of magnesium fluoride, and it is found that 0.0080 g of the compound will dissolve in 500.0 mL water. What is K_{sp} for MgF_2?

Determination of Solubility

Solubility product constant values are known for a large number of compounds, so in the majority of cases, they can be obtained from reference sources. Therefore, a more common calculation is to find solubility from known K_{sp} values. The procedure for determining solubility from the K_{sp} value is as follows:

1. Write the balanced equation representing the process of dissolution.
2. Write the K_{sp} expression, based on the equation from Step 1.
3. Assign the algebraic variable x to one of the ions on the right side of the equilibrium equation that has the same stoichiometric coefficient as the solid on the left side of the equation.
4. Assign algebraic variables for the other ions on the right side of the equilibrium equation in terms of their relationship to the ion from Step 3.

5. Substitute the algebraic variables into the expression for K_{sp}, solve for x, and determine the solubility in the units requested in the statement of the problem.

EXAMPLE 21.2

Determine the solubility of lead(II) bromide, in g/100 cm^3, given that one reference source gives its K_{sp} as 9×10^{-6}.

SOLUTION

The first two steps are to write the equation and the K_{sp} expression:

$$PbBr_2(s) \rightleftharpoons Pb^{2+}(aq) + 2\,Br^-(aq) \qquad K_{sp} = [Pb^{2+}]\,[Br^-]^2$$

We want to know the solubility of $PbBr_2(s)$. There is a 1:1 stoichiometric relationship between $PbBr_2(s)$ and $Pb^{2+}(aq)$, so if we find the $[Pb^{2+}]$, we will have the solubility of $PbBr_2(s)$. Thus, we let $x = [Pb^{2+}]$. Since the coefficient on $Br^-(aq)$ is twice that on $Pb^{2+}(aq)$, it follows that $[Br^-] = 2x$. This completes Steps 3 and 4. Now we substitute and solve:

$$K_{sp} = [Pb^{2+}]\,[Br^-]^2 = 9 \times 10^{-6} = (x)\,(2x)^2$$

$$9 \times 10^{-6} = 4x^3$$

$$x = 0.013 \text{ mol/L} = [Pb^{2+}] = \text{solubility of } PbBr_2(s)$$

To complete the problem, we need to convert moles per liter to grams per 100 cm^3:

$$\frac{0.013 \text{ mol } PbBr_2}{L} \times \frac{367.0 \text{ g } PbBr_2}{\text{mol } PbBr_2} \times \frac{0.1\text{ L}}{100 \text{ cm}^3} = 0.5 \text{ g } PbBr_2/100 \text{ cm}^3$$

Self-Test 2

1. The solubility product constant for silver sulfate is 1.4×10^{-5} at 25°C. What is the solubility (g/100 mL water) of the compound?

The Ion Product

The *ion product*, IP, for a slightly soluble salt has the same form as the expression for the solubility product constant, but it is free from the restriction of applying only to an equilibrium situation. The ion product is valid whether an equilibrium situation does or does not exist, whereas K_{sp}, an equilibrium constant, refers just to an equilibrium situation. Returning to our silver chloride example, we have

$$AgCl(s) \rightleftharpoons Ag^+(aq) + Cl^-(aq) \qquad K_{sp} = [Ag^+][Cl^-] \qquad IP = [Ag^+][Cl^-]$$

The ion product is particularly useful in predicting whether precipitation will occur when solutions contain ions that can form a slightly soluble salt.

If we allow a saturated solution of silver chloride to come to equilibrium, IP will be equal to K_{sp}:

$$IP = [Ag^+][Cl^-] = K_{sp}$$

In other words, the ion product and the equilibrium constant are the same for an equilibrium situation. If, however, we were to begin with a system not at equilibrium and then give that system time to come to equilibrium, the equilibrium will occur at concentrations such that $IP = K_{sp}$.

What about when the system is not at equilibrium? First, let's consider the case when IP is less than K_{sp} ($IP < K_{sp}$). In this case, the solution is unsaturated because there is not a great enough concentration of ions for crystallization to exceed dissolution. No precipitate will form.

The other nonequilibrium situation occurs when the ion product is greater than the equilibrium constant ($IP > K_{sp}$). In this case, the concentration of the aqueous ions in solution is greater than the equilibrium concentration. Therefore, for the system to come to equilibrium, the solution ion concentrations must decrease. Solid will precipitate until $IP = K_{sp}$. Thus, if solutions are combined such that $IP > K_{sp}$, precipitation will occur.

In sum, if solutions are combined such that IP is greater than K_{sp}, the slightly soluble salt will precipitate. If the ion product does not exceed K_{sp}, no precipitation will occur.

EXAMPLE 21.3

In Example 21.1, we determined that the K_{sp} of silver bromide is 4.8×10^{-13}. If 50.0 mL of 0.010 M silver nitrate and 100.0 mL 0.0020 M potassium bromide are combined, will a precipitate form?

SOLUTION

As with all problems involving slightly soluble salts, we start with the equilibrium equation and the K_{sp} expression. This time, however, the reaction for the potential precipitation of the slightly soluble salt is not immediately obvious. Let's consider what ions we have in solution: $Ag^+(aq)$ and $NO_3^-(aq)$ from the silver nitrate solution and $K^+(aq)$ and $Br^-(aq)$ from the potassium bromide solution. A double replacement reaction is possible, forming AgBr and KNO_3. From the solubility rules, we know that potassium and nitrate salts are soluble and that bromide salts are generally soluble, but AgBr is an exception. Thus, the precipitation or dissolution equilibrium is

$$AgBr(s) \rightleftharpoons Ag^+(aq) + Br^-(aq) \quad K_{sp} = [Ag^+][Br^-] = 4.8 \times 10^{-13}$$

The ion product expression follows from the K_{sp} expression:

$$IP = [Ag^+][Br^-]$$

Next, we need the concentrations of the silver and bromide ions. When the two solutions are combined, each original solution is diluted; thus, we use the formula for calculating a dilution, $M_cV_c = M_dV_d$, where M is molarity and V is volume, and the subscripts c and d stand for "concentrated" and "dilute," respectively.

For silver ion,

$$M_d = \frac{M_c \times V_c}{V_d} = \frac{0.010 \text{ M } Ag^+ \times 50.0 \text{ mL}}{150.0 \text{ mL}} = 0.0033 \text{ M } Ag^+$$

For bromide ion,

$$M_d = \frac{M_c \times V_c}{V_d} = \frac{0.0020 \text{ M } Br^- \times 100.0 \text{ mL}}{150.0 \text{ mL}} = 0.0013 \text{ M } Br^-$$

Note that the total volume V_d of solution after the two solutions are combined is the sum of the volumes of the original solutions: 50.0 mL + 100.0 mL = 150.0 mL.

Now we can calculate the ion product and compare it with the K_{sp} value:

$$IP = [Ag^+][Br^-] = (0.0033)(0.0013) = 4.3 \times 10^{-6} > 4.8 \times 10^{-13} = K_{sp}$$

Since $IP > K_{sp}$, we predict that a precipitate of AgBr will form when the solutions are combined.

Self-Test 3

1. A solution that is 0.0015 M in barium chloride is added to a solution that is 0.0010 M in sodium sulfate. Equal volumes of both solutions were combined. Given that K_{sp} = 1.1×10^{-10} for barium sulfate, will precipitation occur or will the solution remain unsaturated?

Workshop: Solubility Equilibria

1. A number of factors can complicate calculations of equilibria involving slightly soluble salts, and these complications form the theme of this workshop. In this question, we want you to consider how the solubility of a salt is affected when it is dissolved in pure water, versus when it is dissolved in a solution that contains one of the ions found in the slightly soluble salt.

   $K_{sp} = 1.4 \times 10^{-10}$ for copper(II) carbonate.

   a. What is the solubility of copper(II) carbonate in pure water? Answer in g/100 mL.

   b. Use Le Chatelier's principle to predict the shift that would occur in the copper(II) carbonate equilibrium if carbonate ion were added to the system.

   c. What is the solubility of copper(II) carbonate in 0.0010 M sodium carbonate?

   d. How does your prediction from Part (b) correspond with the numerical results from Parts a and c?

2. The solubility of slightly soluble salts can be affected by the formation of complex ions. A complex ion is composed of a central positively charged ion and attached electron-pair donor species. In a solution containing silver and chloride ions, for example, the dichloroargentate(I) ion forms:

$$Ag^+(aq) + 2\ Cl^-(aq) \rightleftharpoons AgCl_2^-(aq) \quad K = 1 \times 10^6$$

Of course, the equilibrium between the slightly soluble salt and its ions occurs simultaneously in the solution:

$$AgCl(s) \rightleftharpoons Ag^+(aq) + Cl^-(aq) \quad K_{sp} = 1.8 \times 10^{-10}$$

How will the solubility of a slightly soluble salt be affected if its ions form a complex ion? To answer this question, perform the following calculations:

a. Determine the g/100-mL solubility of silver chloride, neglecting the effect of the complex ion formation.

b. Add the two equations to get the net overall reaction, and calculate the resulting equilibrium constant.

c. Determine the g/100-mL solubility for the overall process.

d. Compare the solubilities. What effect does complex ion formation have on the solubility of the salt?

3. Consider these qualitative experimental results:

(i) Very dilute solutions of $Ba(NO_3)_2(aq)$ and $Na_2CO_3(aq)$ are poured into a large beaker half-filled with pure water. A hazy cloud forms.

(ii) Very dilute solutions of $Ba(NO_3)_2(aq)$ and $Na_2CO_3(aq)$ are poured into a large beaker half-filled with KCl(aq) solution. The solution remains clear.

a. What causes the hazy cloud in Experiment i?

b. Experiment ii is an example of a phenomenon called the *salt effect*, wherein the solubility of a slightly soluble salt is increased by the presence of ions in the solution that are not common in the slightly soluble salt. In this case, potassium ions are attracted to the carbonate ions in the solution. Similarly, chloride ions are attracted to the barium ions. These attractions keep the barium and carbonate ions from combining to form a precipitate, as long as they are in low concentration relative to their solubility limit. Sketch a submicroscopic-level diagram of this effect, and explain how your sketch illustrates the *salt effect.*[*]

[*]If your instructor has introduced the *activity* concept, discuss how the activity coefficients of the ions play a role in this situation.

4. If we use the solubility product constant to predict the solubility of silver bromide in pure water, the result of the theoretically based calculation matches the experimental result. By contrast, the experimentally determined solubility of silver carbonate does not match the solubility determined from a K_{sp} calculation.

a. $K_{sp} = 8.1 \times 10^{-12}$ for silver carbonate. Use this value to determine the theoretical solubility (g/100 mL) in pure water.

b. Bromide ion is the conjugate base of a strong acid; carbonate ion is the conjugate base of a weak acid. How do these facts correlate with the theoretical and experimental results for silver bromide and silver carbonate? Do you expect silver carbonate to be more or less soluble than the K_{sp} calculation would predict? Why?

c. Assume that the carbonate ion reacts to form its conjugate acid in solution:

$$CO_3^{2-}(aq) + H_2O(\ell) \rightleftharpoons HCO_3^{-}(aq) + OH^{-}(aq) \quad K = 1.8 \times 10^{-4}$$

Calculate the solubility (g/100 mL) that should result from the combination of the two equilibrium systems.

5. Consider again dissolving silver bromide and silver carbonate. If your objective was to get each of these slightly soluble compounds to dissolve, and you had available concentrated solutions of hydrochloric acid and hydrobromic acid, what would be the effect on each of the salts? Qualitatively consider all four possible combinations.

6. Another complicating factor in predicting solubilities from K_{sp} values involves the formation of ion pairs in solution. An *ion pair* is an aqueous pair of ions that acts like a single particle in solution. The pair is formed by oppositely charged ions in the solution. When solid calcium sulfate is placed in pure water and allowed to come to equilibrium, the expected dissolution occurs to a small extent:

$$CaSO_4(s) \rightleftharpoons Ca^{2+}(aq) + SO_4^{2-}(aq)$$

In addition, the solid dissolves to form ion pairs:

$$CaSO_4(s) \rightleftharpoons Ca^{2+}SO_4^{2-}(aq)$$

In this reaction, $Ca^{2+}SO_4^{2-}(aq)$ represents the ion pair formed from the calcium ion and the sulfate ion.

The ions in solution can also form ion pairs:

$$Ca^{2+}(aq) + SO_4^{2-}(aq) \rightleftharpoons Ca^{2+}SO_4^{2-}(aq)$$

All three equilibria must be considered in order to understand the solubility of calcium sulfate.

a. Does ion-pair formation increase or decrease the solubility of a slightly soluble salt? Explain.

b. Ion-pair formation is greatest for highly charged ions and for small ions. List a few slightly soluble salts that you would expect to have a relatively large ion-pair effect and a few for which you would expect this effect to be relatively small.

7. In Questions 1 through 6, your team considered and discussed the major complicating factors in the relationship between theoretical and experimental solubilities. To wrap up this workshop, (a) generate a list of these factors, (b) give a brief explanation of each, and (c) give the direction in which solubility shifts compared with a situation in which the complicating factor does not exist.

Entropy Concepts

> *"A sweet disorder in the dress*
> *Kindles in clothes a wantonness"*
> *Robert Herrick*

The universe is made up of matter and energy. You have spent a great deal of time studying the transformations of matter thus far in your chemistry course, but what about energy transformations? Amazingly, only a few simple scientific laws are needed to form the foundation of our understanding of how energy changes from one form to another. Collectively, these laws are known as the laws of thermodynamics.

The First Law of Thermodynamics states that, for an ordinary (nonnuclear) physical or chemical change, energy is neither created nor destroyed during the change. Instances of this law are often not immediately apparent, because energy changes can be difficult to detect with our human senses, but when we use sufficiently sensitive equipment, we invariably find that the total amount of energy both before and after a change are always the same. If we look at the First Law on a universal scale, it would be stated by saying that the total amount of energy in the universe is constant.

Spontaneous Changes

The focus of our studies in this unit will be on the Second and Third Laws of Thermodynamics and their applications. Before we discuss these laws, we must introduce the concept of spontaneity.

A spontaneous change is a change that occurs without outside intervention. If you touch a hot burner on a stove top, your fingers will get burned. The heat transfers spontaneously from the burner to your finger. If you place a water-filled container outside on a frigid January morning, the water changes spontaneously to ice. In this case, the heat moves spontaneously from the water to the colder air. Notice that, in your real-life experiences, the direction of spontaneous changes such as these is invariably the same: Heat always "flows" from the object at the higher temperature to the object at the lower temperature.

The Second Law of Thermodynamics

The fact that chemical and physical changes have a natural, spontaneous direction in which they occur begins to hint at the inadequacy of the First Law of Thermodynamics standing alone as a method for understanding energy changes. The total amount of energy involved in a change is indeed constant, but there is also a natural direction for this energy flow. This direction is given by the Second Law.

The Second Law of Thermodynamics states that changes occur in such a direction that the disorder, or randomness, of the universe increases. Scientists use the term *entropy* to refer to disorder at the molecular level. Thus, another way of stating the Second Law is to say that the entropy of the universe increases for all spontaneous changes.

A Statistical View of the Second Law

The Second Law of Thermodynamics can also be described mathematically. The Austrian physicist Ludwig Boltzmann (1844–1906) showed that the entropy S of a collection of particles is proportional to the natural logarithm of the number W of different particle arrangements that result in the same macroscopic situation:

$$S \propto \ln W$$

Introducing a proportionality constant k, we have a mathematical statement of the Second Law:

$$S = k \ln W$$

The proportionality constant in this equation, known as the *Boltzmann constant*, is the gas constant per molecule:

$$k = \frac{R}{N} = \frac{8.315 \text{ J/mol} \cdot \text{K}}{6.022 \times 10^{23} \text{/mol}} = 1.381 \times 10^{-23} \text{ J/K}$$

In order to gain a better understanding of the mathematical approach to the Second Law, let's consider a universe composed of only four atoms and nine places in which they can reside. We can think of the nine locations in terms of a grid:

Our first atom can reside in any of the nine locations. The second atom then has only eight possible locations. The third and fourth atoms have seven and six potential locations, respectively. The number of possible arrangements for our four atoms is, then,

$$W = 9 \times 8 \times 7 \times 6 = 3024$$

This is true, however, only if the four atoms are unique or different from one another. If all four atoms are the same, the number of possible arrangements must be reduced. The arrangements

a	b	
	c	
		d

and

b	c	
	d	
		a

are the same if all four atoms are identical:

x	x	
	x	
		x

Therefore, we need a correction in W for the four identical atoms. Since the label a could be placed on any of the four atoms, and the label b could be placed on any of the three remaining atoms, etc., we need to reduce W by

$$4 \times 3 \times 2 \times 1 = 24$$

which results in

$$W = \frac{9 \times 8 \times 7 \times 6}{4 \times 3 \times 2 \times 1} = 126$$

The entropy of our four-atom, nine-location universe would therefore be

$$S = k \ln W = (1.381 \times 10^{-23}\ J/K) \times \ln 126 = 6.679 \times 10^{-23}\ J/K$$

How much entropy is this? One extreme that we can use for comparison is the situation in which there is only one way to arrange the atoms, when W = 1. In this case,

$$S = k \ln W = k \ln 1 = 0$$

Since there is only one arrangement that leads to a perfect crystalline solid at the lowest temperature possible, W = 1 and S = 0. This result leads to the Third Law of Thermodynamics, which says that the entropy of a pure, perfect crystalline solid at 0 K is zero.

Self-Test 1

1. Write a brief, succinct statement to summarize the laws of thermodynamics.

2. Calculate the entropy of a three-atom, six-location universe in which all three atoms are identical.

The entropies of a few common substances at 25°C and 1 atm are as follows:

Solid aluminum	$S° = 28.32$ J/mol K
Solid carbon (diamond)	$S° = 2.43$ J/mol K
Liquid water	$S° = 69.91$ J/mol K
Liquid mercury	$S° = 77.40$ J/mol K
Gaseous nitrogen	$S° = 191.50$ J/mol K
Gaseous oxygen	$S° = 205.0$ J/mol K

The degree sign associated with the symbol for entropy indicates standard thermodynamic conditions.

Notice that the values for the entropies in the preceding list all far exceed the value of entropy calculated for our four-atom, nine-location universe. This difference is logical, given that those entropies are for a mole of atoms, rather than just four, and that that mole of atoms is in the real universe, which has many, many locations.

Notice also another pattern in the entropies in the list: The entropies of the solids are lower than those of the liquids, which in turn are lower than those of the gases. This ordering correlates with our understanding of kinetic molecular theory and the behavior of particles at the molecular level: Of the three common states of matter, the solid state is the most orderly at the particle level, the liquid state is more disordered, and the gaseous state is even more disordered. The numeric values of entropy reflect this increasing disorder.

Self-Test 2

1. Consider each of the following pairs of substances, predict which will have the greatest entropy, and give a brief explanation justifying your choice:

   a. Solid dry ice (solid carbon dioxide) or gaseous carbon dioxide

   b. Steam or liquid water

   c. Liquid nitrogen or gaseous nitrogen

Workshop: Entropy Concepts

1. Divide your workshop group into two groups of four if possible. Each member of the subgroup needs one coin.

In this problem, we will consider a universe that contains four different, distinguishable atoms and consists of two locations. Our ultimate goal is to determine W for this universe.

To begin, let's investigate the likelihood of each possible arrangement for this universe. Each person will represent one of the four atoms. If the coin flip results in heads, that atom is in location H. If the flip results in tails, that atom is in location T. Remember, H and T are the only possible locations in this made-up universe.

Assign the numbers 1 through 4 to each person in the subgroup. Flip a total of 100 coins and record the location of each atom (either H or T) in the table below.

Trial Number	Atom 1	Atom 2	Atom 3	Atom 4	Trial Number	Atom 1	Atom 2	Atom 3	Atom 4
1					51				
2					52				
3					53				
4					54				
5					55				
6					56				
7					57				
8					58				
9					59				
10					60				
11					61				
12					62				
13					63				
14					64				
15					65				
16					66				
17					67				
18					68				
19					69				
20					70				
21					71				
22					72				
23					73				
24					74				
25					75				
26					76				
27					77				

28					78				
29					79				
30					80				
31					81				
32					82				
33					83				
34					84				
35					85				
36					86				
37					87				
38					88				
39					89				
40					90				
41					91				
42					92				
43					93				
44					94				
45					95				
46					96				
47					97				
48					98				
49					99				
50					100				

a. Summarize your results. How many different arrangements were possible? How many different arrangements actually occurred? How many times did each arrangement occur? Compare your results with those of the other subgroup.

b. What do the results of this question have to do with the Second Law of Thermodynamics? Explain.

2. Let's revisit the idea of determining W for a four-atom, two-location universe as in Question 1, but this time let's approach the problem from a theoretical point of view. How many possible states are there for this made-up universe? Draw a representation of each possible state. For example, one state would have all four atoms in one location, and it might be represented as

① ② ③ ④	

A second state would have Atoms 1, 2, and 3 on the left and Atom 4 on the right:

① ② ③	④

Complete this question by drawing each of the other possible states. How does this result compare with the one from Question 1?

3. Calculate the numeric value of entropy for the made-up universe of Questions 1 and 2. How does this value compare with that of real substances? Explain.

4. Consider each of the entropy changes that follow, and decide whether ΔS will be positive or negative by considering the relative entropies of the reactants and products. Explain your answer to the group.

a. $H_2O(\ell) \rightarrow H_2O(g)$

b. $CaCO_3(s) \rightarrow CaO(s) + CO_2(g)$

c. $N_2(g) + 3\,Cl_2(g) \rightarrow 2\,NCl_3(g)$

d. $Ba(NO_3)_2(aq) + H_2SO_4(aq) \rightarrow BaSO_4(s) + 2\,HNO_3(aq)$

Enthalpy, Entropy, and Free-Energy Calculations

> *"The energies of our system will decay, the glory of the sun will be dimmed, and the earth, tideless and inert, will no longer tolerate the race which has for a moment disturbed its solitude. Man will go down into the pit, and all his thoughts will perish."*
> *A. J. Balfour*

Can we predict whether a reaction will occur? This is a question that can be answered by applying the principles of chemical thermodynamics—the study of the energy relationships associated with chemical reactions. The term we use to label a reaction that proceeds without a continual input of energy is *spontaneous.* A spontaneous reaction is a reaction that will occur all by itself, once it has been given a small amount of energy so that it can get started. The burning of paper, for example, is a spontaneous reaction. Once you add a little bit of energy, like the heat from a match, the paper continues to burn without any outside help, until there is no more paper to burn. In contrast, a *nonspontaneous* reaction is a reaction that will not proceed unless an outside source of energy is used. An example of a nonspontaneous reaction is the decomposition of water into hydrogen and oxygen. If we add energy to water, the water may begin to decompose (if the amount of energy is great enough), but the decomposition will stop as soon as the energy source is cut off. In this unit, we focus on performing calculations that will allow us to predict the spontaneity of chemical reactions.

Free Energy

J. Willard Gibbs (1839–1903) can be considered one of the founding fathers of the field of chemical thermodynamics. He introduced a quantity known as the *Gibbs free energy* G, which represents the amount of energy available in a chemical system that can do useful work. If we are to consider a chemical change, we are interested in the change in free energy, or ΔG. Thus, ΔG is a measure of the amount of energy in a chemical change that is free to do work on another physical or chemical system.

The most notable aspect of the ΔG concept is that its sign allows us to predict the spontaneity of a chemical reaction under conditions of constant temperature and pressure:

1. If ΔG is negative, the reaction is spontaneous.
2. If ΔG is positive, the reaction is nonspontaneous.
3. If ΔG is zero, the reacting system is at equilibrium, and there will be no change in the reaction on the macroscopic level.

Standard state values of ΔG, symbolized as $\Delta G°$, are commonly found in tables of thermodynamic quantities. Recall that the thermodynamic standard state conditions are 25°C, 1 atm pressure for gases, and 1-M concentrations for solutions. The calculation of $\Delta G°$ for a reaction is given by

$$\Delta G° = \Sigma n \Delta G_f°{}_{\text{products}} - \Sigma n \Delta G_f°{}_{\text{reactants}} \qquad \textbf{(23.1)}$$

where $\Delta G_f°$ is the standard free energy of formation—that is, the free-energy change that occurs when 1 mole of a compound is formed from elements in their standard states. Note the similarity of this equation to the equation used to calculate $\Delta H°$ for a reaction, which was introduced earlier in your study of chemistry.

EXAMPLE 23.1

Calculate the free-energy change for the complete combustion of 1 mole of methane [$CH_4(g)$], the main component of natural gas. Is this reaction spontaneous?

SOLUTION

We begin by writing the equation that represents the reaction. Recall that "complete combustion," or burning, is a reaction with oxygen from the atmosphere, forming carbon dioxide and water:

$$CH_4(g) + 2\,O_2(g) \rightarrow CO_2(g) + 2\,H_2O(\ell)$$

Now we apply Equation 23.1 and then consult a table of thermodynamic values:

$$\begin{aligned}\Delta G° &= (1\text{ mol})[\Delta G_f° \text{ for } CO_2(g)] + (2\text{ mol})[\Delta G_f° \text{ for } H_2O(\ell)] - \\ &\quad (1\text{ mol})[\Delta G_f° \text{ for } CH_4(g)] - (2\text{ mol})[\Delta G_f° \text{ for } O_2(g)] = \\ &(1\text{ mol})(-394.4\text{ kJ/mol}) + (2\text{ mol})(-237.0\text{ kJ/mol}) - (1\text{ mol})(-50.8\text{ kJ/mol}) - (2\text{ mol})(0) = \\ \Delta G° &= -817.6\text{ kJ}\end{aligned}$$

($\Delta G_f°$ for an element in its most stable form under standard conditions is zero.)

The negative value of $\Delta G°$ indicates that the reaction is spontaneous. This matches our experiences in everyday life, where we have seen that natural gas burns spontaneously.

Self-Test 1

1. Solid elemental sulfur can be produced, along with liquid water, by the reaction of hydrogen sulfide and sulfur dioxide gases. Calculate the standard free-energy change for this reaction.

Entropy

The standard-state entropy change, $\Delta S°$ for a reaction can be calculated from data in thermodynamic tables in a manner similar to changes in enthalpy and free energy. $\Delta S°$ for a chemical reaction is

$$\Delta S° = \Sigma nS°_{\text{products}} - \Sigma nS°_{\text{reactants}} \qquad \textbf{(23.2)}$$

A notable difference in $\Delta S°$ values is that we do not use "entropies of formation." This is a result of the Third Law of Thermodynamics, which defines a zero for entropy and thus allows us to calculate absolute entropy values.

EXAMPLE 23.2

Determine the standard entropy change for the decomposition of 1 mole of solid calcium carbonate, forming solid calcium oxide and carbon dioxide gas.

SOLUTION

This is a straightforward application of Equation 23.2, followed by substitution of the appropriate values from a table:

$$CaCO_3(s) \rightarrow CaO(s) + CO_2(g)$$

$\Delta S°$ = (1 mol)[$\Delta S°$ for $CaO(s)$] + (1 mol)[$\Delta S°$ for $CO_2(g)$] –
(1 mol)[$\Delta S°$ for $CaCO_3(s)$] =
(1 mol) (39.8 J/mol K) + (1 mol) (213.7 J/mol K) – (1 mol) (92.9 J/mol K) =
160.6 J/K

Self-Test 2

1. Nitrogen monoxide gas spontaneously decomposes into dinitrogen oxide and nitrogen dioxide gases. What is the standard entropy change for the decomposition of 3.0 mol of nitrogen monoxide?

Free Energy, Enthalpy, and Entropy

By definition, the Gibbs free energy is

$$G \equiv H - TS \quad \textbf{(23.3)}$$

We are interested in the change in free energy associated with chemical reactions, rather than in absolute quantities, so we have

$$\Delta G = \Delta H - \Delta(TS) \quad \textbf{(23.4)}$$

If we consider constant-temperature processes,

$$\Delta G = \Delta H - T\Delta S \quad \textbf{(23.5)}$$

Finally, adding in standard-state conditions, we have

$$\Delta G^\circ = \Delta H^\circ - T\Delta S^\circ \quad \textbf{(23.6)}$$

Equation 23.6, the Gibbs–Helmholtz equation, tells us that the standard free energy change depends on both the change in enthalpy and the change in entropy. We will explore this idea further during the workshop activities; for now, let's see how ΔG° can be calculated from ΔH° and ΔS°.

EXAMPLE 23.3

Calculate $\Delta G°$ for the reaction in Example 23.2—the decomposition of calcium carbonate—from $\Delta H°$ and $\Delta S°$ values.

SOLUTION

We have already calculated $\Delta S°$ for the reaction $CaCO_3(s) \rightarrow CaO(s) + CO_2(g)$ as 160.6 J/K. We can find $\Delta H°$ for the reaction in a similar manner:

$$\Delta H° = (1\ \text{mol})[\Delta H_f°\ \text{for}\ CaO(s)] + (1\ \text{mol})[\Delta H_f°\ \text{for}\ CO_2(g)] - (1\ \text{mol})[\Delta H_f°\ \text{for}\ CaCO_3(s)] = (1\ \text{mol})\ (-635.3\ \text{kJ/mol}) + (1\ \text{mol})\ (-393.5\ \text{kJ/mol}) - (1\ \text{mol})\ (-1207.0\ \text{kJ/mol}) = 178.2\ \text{kJ}$$

Now we use Equation 23.6 to find the value of $\Delta G°$:

$$\Delta G° = \Delta H° - T\Delta S° = 178.2\ \text{kJ} - 298.15\ \text{K} \times \frac{160.6\ \text{J}}{\text{K}} \times \frac{1\ \text{kJ}}{1000\ \text{J}} = 130.3\ \text{kJ}$$

Note that we used 298.15 K, or 25°C, as the value of T. This is the standard thermodynamic temperature. Note also how the value of $\Delta S°$ was in joules per kelvin, while the value of $\Delta H°$ was in kJ. The J $\leftrightarrow$ kJ conversion must be accounted for in the calculation.

Self-Test 3

1. A hypothetical reaction has $\Delta H° = -200.3$ kJ and $\Delta S° = -77.0$ J/K. At what temperature is this reaction spontaneous? Support your answer with the appropriate calculation.

Nonstandard Conditions and Free Energy

Up to this point, we have considered thermodynamic changes only under standard conditions. ΔG at any conditions can be determined by

$$\Delta G = \Delta G° + RT \ln Q \qquad \textbf{(23.7)}$$

where Q is the reaction quotient—the same quantity that was introduced during our study of chemical equilibria.

EXAMPLE 23.4

Consider the reaction of nitrogen monoxide and chlorine to form nitrosyl chloride:

$$2\,NO(g) + Cl_2(g) \rightarrow 2\,NOCl(g)$$

a. Calculate $\Delta G°$ for the reaction.
b. Calculate ΔG when $p_{NO} = 0.30$ atm, $p_{Cl_2} = 0.10$ atm, and $p_{NOCl} = 0.45$ atm.

SOLUTION

a. $\Delta G°$ is found by applying Equation 23.1:

$\Delta G° = (2\text{ mol})[\Delta G_f° \text{ for } NOCl(g)] - (2\text{ mol})[\Delta G_f° \text{ for } NO(g)] -$
$(1\text{ mol})[\Delta G_f° \text{ for } Cl_2(g)] =$
$(2\text{ mol})(66.2\text{ kJ/mol}) - (2\text{ mol})(86.6\text{ kJ/mol}) - (1\text{ mol})(0) = -40.8\text{ kJ}$

b. ΔG at nonstandard conditions (the pressures are *not* 1 atm in this case) is found by applying Equation 23.7: $\Delta G = \Delta G° + RT \ln Q$. Let's begin by calculating Q:

$$Q = \frac{(p_{NOCl})^2}{(p_{NO})(P_{Cl_2})} = \frac{(0.45)^2}{(0.30)^2\,(0.10)} = 23$$

Now we can find ΔG:

$$\Delta G = \Delta G° + RT \ln Q = -40.8\text{ kJ} + \frac{8.315\text{ J}}{\text{mol} \cdot \text{K}} \times 298.15\text{ K} \times \ln 23 \times \frac{1\text{ kJ}}{1000\text{ J}} =$$

-33.0 kJ

Self-Test 4

1. Most of the direct energy needs of a cell are provided by the reaction of adenosine 5'-triphosphate (ATP) to form adenosine 5'-diphosphate (ADP) and hydrogen phosphate ion (P_i):

$$ATP \rightarrow ADP + P_i$$

$\Delta G° = -30.0$ kJ/mol for this reaction. What is ΔG when the concentrations in a cell are [ATP] = 3.2×10^{-3} M, [ADP] = 1.4×10^{-3} M, and [P_i] = 5.0×10^{-3} M and the temperature is 98.6°F?

The Equilibrium Constant and Free Energy

Starting from Equation 23.7,

$$\Delta G = \Delta G° + RT \ln Q \qquad \textbf{(23.7)}$$

and using the facts that, at equilibrium, $Q = K$ and $\Delta G = 0$, we obtain

$$0 = \Delta G° + RT \ln K \qquad \textbf{(23.8)}$$

Rearranging terms to isolate $\Delta G°$ yields

$$\Delta G° = -RT \ln K \qquad \textbf{(23.9)}$$

Equation 23.9 thus gives the relationship between the change in free energy for a reaction and the equilibrium constant.

EXAMPLE 23.5

K_{sp} for the reaction $BaSO_4(s) \rightleftharpoons Ba^{2+}(aq) + SO_4^{2-}(aq)$ is 1.1×10^{-10}. Use thermodynamic data to determine $\Delta G°$ for this reaction, and then calculate K from Equation 23.9. How do the K values compare?

SOLUTION

$\Delta G°$ is found in the usual manner:

$\Delta G°$ = (1 mol)[$\Delta G_f°$ for $Ba^{2+}(aq)$] + (1 mol)[$\Delta G_f°$ for $SO_4^{2-}(aq)$] – (1 mol)[$\Delta G_f°$ for $BaSO_4(s)$] =
(1 mol) (–560.8 kJ/mol) + (1 mol) (–744.5 kJ/mol) – (1 mol) (–1362.3 kJ/mol) = 57.0 kJ

Now we can use Equation 23.9 to find K:

$$\Delta G° = -RT \ln K; \qquad \ln K = \frac{-\Delta G°}{RT}; \qquad K = e^{(-\Delta G°/RT)}$$

Let's get the power of e first:

$$\frac{-\Delta G°}{RT} = -57.0 \text{ kJ} \times \frac{\text{K}}{8.315 \text{ J}} \times \frac{1}{298.15 \text{ K}} \times \frac{1000 \text{ J}}{\text{kJ}} = -23.0$$

↑ assume molar quantities

$$K = e^{-23.0} = 1.0 \times 10^{-10}$$

The K calculated from $\Delta G_f°$ values agrees with the tabulated K_{sp} value to ±1 in the doubtful digit.

Self-Test 5

1. Find the value of K_{sp} for iron(II) hydroxide from your textbook, and then use Equation 23.9 to determine the value of $\Delta G°$ for the dissolution reaction of this slightly soluble solid. How does the value you found compare with the value determined by using $\Delta G_f°$ values?

Workshop: Enthalpy, Entropy, and Free-Energy Calculations

Questions 1–3: Divide your workshop group into three subgroups, and have each subgroup work on one of the three questions that follow. When each subgroup is finished, it should discuss its results with the whole group.

1. Consider the definition of the Gibbs free energy, $G \equiv H - TS$, and the equations that can be derived from this definition. Construct a "truth table" showing all possible combinations of enthalpy and entropy changes for a chemical reaction and the resulting ability to predict the spontaneity of the reaction. Use your textbook to find an example of a reaction that fits each condition in your truth table.

2. Consider the definition of the Gibbs free energy, $G \equiv H - TS$, and the equations that can be derived from this definition. Can a particular chemical reaction be nonspontaneous at one temperature, yet spontaneous at another temperature? If so, what criteria must be satisfied? If not, explain why not.

3. Consider the equation $\Delta G = \Delta G° + RT \ln Q$ that allows us to calculate free-energy changes under nonstandard conditions and the equations that can be derived from that equation. Can the spontaneity of a reversible chemical reaction be determined solely from the equilibrium constant of that reaction? To answer this question, carefully consider the criteria for determining spontaneity and the relationship between the reaction quotient Q and the equilibrium constant K. Explain your answer.

Questions 4–8: Complete each question, using a round-robin approach. Refer to data from the thermodynamic tables in your textbook as necessary to answer the questions.

4. Determine the free-energy change when 1.00 L of ethane [$C_2H_6(g)$] at 25°C and 1.0 atm pressure is completely oxidized.

5. Consider the decomposition of solid ammonium chloride to ammonia and hydrogen chloride gases. What do you predict for the sign of $\Delta S°$? Give a particulate-level explanation for your prediction. Now calculate the value of $\Delta S°$ for the reaction, and compare it with your prediction. Next, consider the decomposition of aqueous ammonium chloride to aqueous ammonia and hydrogen chloride (What is the common name for aqueous hydrogen chloride? What are the major species in its solution?). What do you predict for the sign of $\Delta S°$? Why? Compare your prediction with the calculated value.

6. Consider these thermodynamic values for hypothetical compounds:

Species (state)	ΔH_f° (kJ/mol)	S° (J/mol K)
A(g)	–386.5	177.0
B(g)	–139.9	234.8
Y(g)	33.6	277.1
Z(g)	–295.2	301.3

Is the reaction A + B → Y + Z spontaneous? Is the reaction Y + Z → A + B spontaneous? How do you know?

7. One of the reaction steps for the metabolism of glucose in animals is not spontaneous:

$$\text{2-phosphoglycerate} \rightarrow \text{phosphoenolpyruvate} \quad \Delta G^{\circ'} = 1.7\ \text{kJ/mol}$$

(The prime indicates the biochemistry standard state, which is the same as the chemistry standard state, except that biochemistry uses pH = 7.0 as a condition.) Will this reaction take place in a cell in which [2-phosphoglycerate] = 2.3×10^{-4} M and [phosphoenolpyruvate] = 8.4×10^{-5} M?

8. Complete the blanks in the following statement: Significant quantities of both reactants and products are present at equilibrium for a reversible chemical reaction if $\Delta G°$ for that reaction is between ______ and ______.

Electrochemistry

> *"'Reduce' is derived from Latin words meaning 'to lead back' and if iron, for instance, is observed to turn to rust . . . it is natural to think of rust as being 'led back' to iron."*
> *Isaac Asimov*

Consider two of the most common words used by electrochemists: *reduction* and *oxidation.* These terms originated from early chemists who studied metal ores. According to Isaac Asimov, the term *reduced* is derived from Latin, meaning "to lead back." For instance, when iron ore (solid iron oxides) is combined with coke (a nearly pure form of carbon derived from coal) and heated, a chemical change takes place whereby the carbon atoms from the coke combine with the oxygen atoms in the ore, producing carbon dioxide, and pure iron is left behind. The iron ore is *reduced.* Since the carbon caused the iron ore to be reduced, it is called the *reducing agent.* Carbon has taken on oxygen atoms, so it is said to be *oxidized.*

The modern definitions of the terms *oxidation* and *reduction* are no longer restricted to reactions involving oxygen. Oxidation is the loss of electrons. Reduction is the gain of electrons. One of the most fascinating aspects of oxidation–reduction reactions is the tendency of various substances to gain or lose electrons. If a substance with a powerful tendency to gain electrons is paired with a substance with a tendency to lose electrons, a *battery* or *fuel cell* is created in which the energy of the electron transfer can be tapped. Utilizing the energy of electron-transfer reactions has far-reaching consequences, ranging from powering calculators and laptop computers to converting food to usable energy in living organisms.

Half Reactions

Consider two half reactions— the reduction of copper(II) ions and the oxidation of nickel metal:

$$Cu^{2+}(aq) + 2\,e^- \rightleftharpoons Cu(s) \quad \textbf{(24.1)}$$

$$Ni(s) \rightleftharpoons Ni^{2+}(aq) + e^- \quad \textbf{(24.2)}$$

Combining these two half reactions, we obtain the complete balanced redox reaction:

$$Cu^{2+}(aq) + Ni(s) \rightleftharpoons Cu(s) + Ni^{2+}(aq)$$

Let's consider how this reaction might be carried out in the laboratory. The simplest method is to place a piece of nickel into a solution that contains copper(II) ions, such as copper(II) chloride. This reaction is illustrated in Figure 24.1. The solution originally is blue, which is the color of copper(II) ions in aqueous solution. As the reaction proceeds, nickel atoms transfer electrons to copper(II) ions, and the resulting nickel ions become hydrated, increasing the concentration of $Ni^{2+}(aq)$ in solution. As the copper(II) ion concentration decreases and the nickel ion concentration in solution increases, the color of the solution changes from blue to green, the color of $Ni^{2+}(aq)$. The energy associated with the electron transfer in this experiment is not used for work and is wasted as heat.

By separating the reactants and connecting them with a wire, the electron will do work as it moves from the nickel to the copper. Since the electron will have a different energy at one end of the connecting wire than the other, a potential-energy difference, or *voltage,* exists. The units of voltage are volts. A potential-energy difference of 1 volt exists when 1 coulomb of electron charge does 1 joule of work as it moves from one point to another.

The definition of a volt makes the definition of a coulomb necessary. We can think of a coulomb as the charge on 6.24×10^{18} electrons. You may wonder why a mole of electrons was not chosen to define a coulomb, but, to the chagrin of chemists (and chemistry students alike!), the coulomb was defined before the mole. Don't let the units make you lose sight of the main point, however: The voltage of an electrochemical cell can be measured, and the greater the voltage, the greater is the potential-energy difference for the movement of electrons from anode to cathode.

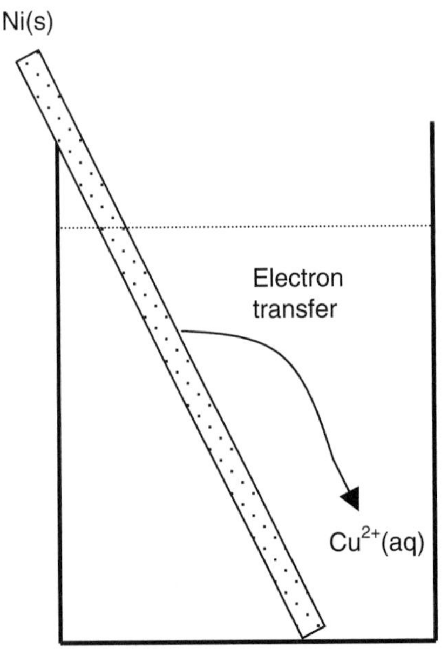

Figure 24.1. A piece of nickel metal is placed in a solution of copper(II) chloride. The solution is originally blue because of the copper(II) ions, but as the reaction proceeds, the solution turns green, as a result of the presence of nickel ions.

Self-Test 1

1. Review the concepts of electron charge, coulombs, and faradays in your textbook before answering the following questions:

   a. How many faradays of charge are in 1 mole of electrons?

   b. How many faradays of charge does a single electron have?

   c. How many coulombs are in 1 mole of electrons?

Consider the cell illustrated in Figure 24.2. When electrons transfer across the cell, a buildup of negative charge occurs on the silver strip, and there is a deficit of electrons on the nickel. After a brief moment, this arrangement will prevent any further flow of electrons, as they will be repulsed by the highly negative electrode. To solve this problem of buildup of charge, a salt bridge is added to the cell to provide an alternative pathway so that ions are free to move from one side to the other.

In an electrochemical cell such as the one illustrated in Figure 24.2, the electrons flow spontaneously, with no external power source. Such a cell is known as a *voltaic,* or *galvanic,* cell. Electrons flow from the anode to the cathode. The salt bridge, filled with an aqueous solution of an ionic compound, such as $NaNO_3(aq)$, allows ions to move to compensate for the buildup of electrons at the cathode. In this cell, the nitrate anions would move from the cathode to the anode.

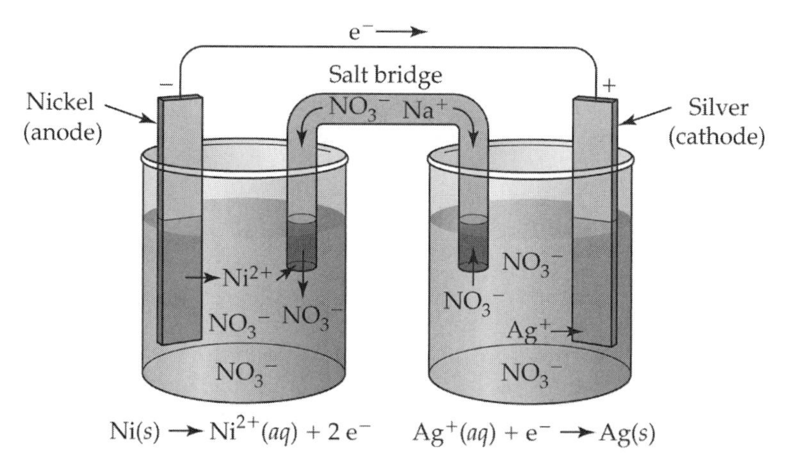

Figure 24.2. An electrochemical cell consisting of one compartment with copper metal submersed in a copper(II) ion-containing solution and the other compartment with nickel metal in a nickel ion-containing solution. The metal strips are connected by a conducting wire, and the solutions are connected by a salt bridge.

If the nickel and silver ion concentrations are 1.00 M in the cell, the potential is 0.55 volt. A useful reference point to use to think about cell potentials is a common alkaline battery, such as that used to power personal stereos and flashlights. This kind of battery has a potential of 1.5 volts. The potential in a cell can be viewed as arising from the relative tendencies of the two half cells to give or take electrons. Each redox half reaction has a half-cell potential. If the potential is measured at standard conditions, it is called the standard half-cell potential, symbolized as E°. For the cell in Figure 24.2, we have the following data:

Half-cell Reaction	Standard Reduction Potential, E° (V)
$Ag^+(aq) + e^- \rightleftharpoons Cu(s)$	0.80
$Ni^{2+}(aq) + 2\,e^- \rightleftharpoons Ni(s)$	–0.25

Standard reduction potentials are determined by constructing a cell and measuring the potential of the half reaction versus the hydrogen half reaction with all solutes at 1.00 M and gases at 1.00 atm. The reaction $2\,H^+(aq) + 2\,e^- \rightarrow H_2(g)$ with 1.00 M $H^+(aq)$ and 1.00 atm $H_2(g)$ is assigned a potential of exactly zero. Therefore, all standard reduction potentials are a measure of the tendency of that half reaction to gain or lose an electron relative to the standard hydrogen electrode.

The total potential for a cell is the sum of the potentials of the half cells. The half-cell reaction for the reduction of the silver ion is exactly as just listed. However, the spontaneous half-cell reaction for Ni is the opposite of what is listed. In general,

$$E^\circ_{total} = E^\circ_{cathode} - E^\circ_{anode}{}^{*} \quad \textbf{(24.4)}$$

Self-Test 2

1. Select two standard half reactions for metal ion reductions from a table of standard half-cell potentials (other than silver or nickel).

   a. Which substance is the reducing agent, and which substance is reduced?

   b. Write a complete oxidation–reduction equation from the two half reactions.

*Some textbooks give this equation as its equivalent, E°cell = E°oxidation + E°reduction.

c. Draw and label a sketch of the voltaic (galvanic) cell that results.

d. Calculate the standard potential for the reaction.

e. Determine the potential when the concentration of the reducing agent is 1.0×10^{-4} M and the concentration of the oxidizing agent is 1.0 M.

The Nernst Equation

From a table of standard potentials, you can calculate the voltage for a cell composed of any combination of half reactions, as long as the cell is operating under standard conditions. When the concentrations are nonstandard, the Nernst equation, named after the German chemist Walther Nernst (1864–1941), allows the voltage to be determined:

$$E = E^\circ - \frac{0.0592}{n} \log Q \qquad \textbf{(24.5)}$$

This equation is valid when potentials are measured in volts and the temperature is 25°C. Q is called the *reaction quotient* and is set up just like an equilibrium constant expression. When a solid or gas is present, its concentration term is set to unity. Dissolved solute concentrations are expressed in moles per liter. The Nernst equation for the silver–nickel cell in Figure 24.2:

$$E = 0.57 - \frac{0.0592}{n} \log \left(\frac{[Ni^{2+}]}{[Ag^{+}]}\right) \qquad \textbf{(24.6)}$$

Electrolytic Cells

To this point, we have considered only voltaic (galvanic) cells, in which a chemical reaction is used to create a potential that can do work. The reverse situation, in which a potential is used to cause a reaction to occur, is also important. Cells that operate in this manner are called *electrolytic cells*. A common example is the recharging of a rechargeable battery, such as that found in cars. The spontaneous reaction that occurs in a lead automobile battery is

$$Pb(s) + PbO_2(s) + 2\,SO_4^{2-}(aq) + 4\,H^+(aq) \rightleftharpoons 2\,PbSO_4(s) + 2\,H_2O(\ell) \qquad \textbf{(24.7)}$$

When the battery is fully discharged, which means that the voltaic reaction has essentially gone to completion, the battery can be recharged by applying a voltage of 2 V per cell and switching the reactions at the anode and cathode. Electrons naturally flow from plus to minus. The electrode marked plus is the cathode, and the electron marked minus is the anode. When the external voltage source is attached to the battery, the plus lead is attached to the anode and the minus to the cathode. This causes the external voltage to be applied against the spontaneous flow of the current in the cell, which regenerates the original reactants.

Self-Test 3

1. a. What are the differences among voltaic, galvanic, and electrolytic cells?

   b. Distinguish between a reducing agent and a substance that becomes reduced.

Workshop: Electrochemistry

Use the following table of standard reduction potentials as necessary in answering the questions in this workshop:

Half-cell Reaction	Standard Reduction Potential, E° (V)
$K^+(aq) + e^- \rightleftharpoons K(s)$	–2.92
$Al^{3+}(aq) + 3\ e^- \rightleftharpoons Al(s)$	–1.69
$2\ NO_3^-(aq) + 2\ H_2O(\ell) + 2\ e^- \rightleftharpoons N_2O_4(g) + 4\ OH^-(aq)$	–0.85
$2\ H_2O(\ell) + 2\ e^- \rightleftharpoons H_2(g) + 2\ OH^-(aq)$	–0.83
$Zn^{2+}(aq) + 2\ e^- \rightleftharpoons Zn(s)$	–0.76
$C_4H_4O_2(aq) + 2\ H^+(aq) + 2\ e^- \rightleftharpoons C_4H_6O_2(aq)$	–0.15
$2\ H^+(aq) + 2\ e^- \rightleftharpoons H_2(g)$	0.00
$AgBr(s) + e^- \rightleftharpoons Ag(s) + Br^-(aq)$	0.071
$Cu^{2+}(aq) + e^- \rightleftharpoons Cu^+(aq)$	0.16
$Cu^+(aq) + e^- \rightleftharpoons Cu(s)$	0.51
$Ag^+(aq) + e^- \rightleftharpoons Ag(s)$	0.80
$O_2(g) + 4\ H^+(aq) + 4\ e^- \rightleftharpoons 2\ H_2O(\ell)$	1.23

1. Answer the questions that follow without using a calculator. In each case, explain the reasoning you used to arrive at your answer.

   a. Which is the stronger oxidizing agent, $Ag^+(aq)$ or $O_2(g)$?

   b. Which is the stronger reducing agent, Al(s) or Zn(s)?

   c. What is the potential for the half reaction $Ag(s) \rightleftharpoons Ag^+(aq) + e^-$?

   d. What is the potential for the half reaction $\frac{1}{2}\ O_2(g) + 2\ H^+(aq) + 2\ e^- \rightleftharpoons H_2O(\ell)$?

2. Recall that $\Delta G°$ is a state function and therefore path independent. The relationship between $\Delta G°$ and $E°$ is given by $\Delta G° = -nFE°$, where n is the number of moles of electrons transferred.

Given two half reactions with potentials $E°_1$ and $E°_2$,

$$\Delta G°_1 + \Delta G°_2 = -nFE°_1 - nFE°_2$$

a. What is the free-energy change for the reduction of 0.50 mol of oxygen? For 1.00 mol of oxygen?

b. Determine the potential and the free-energy change for the following reactions:
   i. solid silver and aqueous hydrogen ion
   ii. liquid water and potassium metal
   iii. aqueous hydrogen ion and potassium metal

c. Will zinc metal spontaneously react with either water or 1.00 M HCl? Explain.

3. a. Write the Nernst equation for the reaction of $Cu^{2+}(aq)$ in the cathodic cell and $Ni^{2+}(aq)$ in the anodic cell.

b. What is the potential of a cell with 0.10 M $Cu^{2+}(aq)$ in the cathodic cell and 0.010 M $Ni^{2+}(aq)$ in the anodic cell?

4. Write the Nernst equation for the reaction $AgBr(s) \rightleftharpoons Ag^{+}(aq) + Br^{-}(aq)$.

5. When the potential for a cell is zero, there is no net tendency for electrons to flow. At equilibrium, the Nernst equation becomes

$$0 = E° - \frac{0.0592}{n} \log K_{eq}$$

a. Why is K_{eq} used in place of the reaction quotient?

b. Determine the value of the equilibrium constant for the reaction of potassium metal and water.

c. Calculate the equilibrium constant for the reaction $Cu(s) + Cu^{2+}(aq) \rightleftharpoons 2\ Cu^{+}(aq)$.

d. Once we have K_{eq} from the standard potential, E°, we can answer other questions about chemical reactions. Given the initial conditions illustrated in the following diagram, what will the final conditions be?

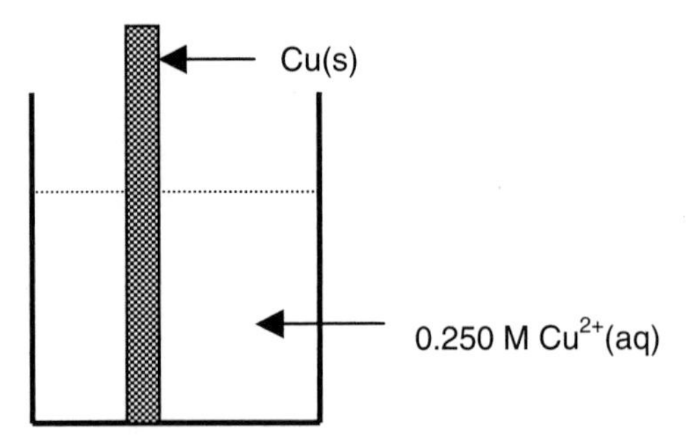

6. An electrolytic cell passes 2.0×10^4 coulombs of charge in reducing aluminum ion to aluminum metal. How many grams of aluminum metal are produced?

7. Although it is not widely discussed, normal physiological redox potential is about –0.1 V. This suggests that the human body maintains a very slightly reducing environment and that there is some evidence that redox potential is related to human health.

Living organisms use redox reactions to transfer electrons. Ubiquinones, also called coenzyme Q, are abundant in biological systems and serve as electron-transport molecules. The reduction half reaction for coenzyme Q (abbreviated Q) is

$$Q(aq) + 2\,H^+(aq) + 2\,e^- \rightleftharpoons QH_2(aq)$$

Since biochemical systems have pH values near 7, biochemists use pH = 7 as their standard state. The standard reduction potential, $E^{\circ\prime}$, where the prime indicates the biochemical standard state, for the coenzyme Q reduction half reaction is 0.04 V.

a. Write the Nernst equation for this reaction, coupled with the nickel half reaction as the anodic reaction.

b. What is the biochemical standard potential for the given reaction?

c. What is the standard potential for the given reaction? [*Hint*: What is the standard-state concentration of $H^+(aq)$?]

8. A chemical research team reports, on the one hand, that the reduction of a certain cobalt compound has a potential of –1.90 V and, on the other hand, that the compound is not likely to be reduced under normal physiological conditions. Explain.

9. A report in the August 3, 1998, issue of *Chemical and Engineering News* talked about the zinc air fuel cell. An electrochemical fuel cell is a galvanic cell in which one of the components needs to be constantly replenished. Oxygen is the component that must be replenished in the zinc air fuel cell. The reaction at the cathode is

$$\frac{1}{2}\,O_2 + H_2O + 2\,e^- \rightarrow 2\,OH^- \qquad E^\circ = 0.401\ V$$

and the reaction at the anode is

$$Zn + 2\,OH^- \rightarrow ZnO + H_2O + 2\,e^- \qquad E^\circ = 1.260\ V$$

a. What is the standard potential for this cell?

b. When the cell is recharged, the reaction runs in reverse. How many moles of oxygen are released when 2.00 g ZnO is reacted while the cell is recharging?

c. What STP volume of oxygen is released when 2.00 g ZnO is reacted while the cell is recharging?

d. How many coulombs of charge are needed for the reaction of 2.00 g ZnO?